CONTENTS

INTRODUCTION by Jonathan Downes	5
A TRIO OF MYSTERY CATS AT LONDON ZOO by Dr Karl P.N.Shuker	7
THE STRANGE ANIMALS OF SWANSEA BAY by Oll Lewis	15
PLANET OF THE APE MEN by Nick Redfern	29
TOWARDS A POSSIBLE CAUDATA IDENTITY FOR THE MONGOLIAN DEATH WORM: *Introducing the 'plausibility method' for identity theory formation amongst lesser known cryptids* by Michael A. Woodley	39
CATS AROUND THE CAPITAL by Neil Arnold	49
ZOOLOGICAL CURIOSITIES FROM *HARDWICKE'S SCIENCE GOSSIP* PART ONE - 1865-7 BY Richard Muirhead	99
ON THE TRACK OF ORANG PENDEK? by Nick Molloy	105
MADNESS, MONSTERS AND MORAR by Lisa Dowley	132
SOME NEW ZEALAND CRYPTIDS by Tony Lucas	145
SINGING MICE by Jonathan Downes	169
ARCHIVE ARTICLES FROM THE 1930s ON SINGING MICE	179
CFZ ANNUAL REPORT 2007	183

Jonathan Downes
(Director, Centre for Fortean Zoology)

-CFZ YEARBOOK 2007-

Dear Friends,

It is with great pleasure that I find myself in the happy position of writing the introduction to this, the tenth of the CFZ Yearbooks. Those of you with even the most rudimentary grasp of basic arithmetic will be counting on your fingers at this point, and realising that more than ten years have gone by since the first volume in 1996, but here I must blush and remind you that the two years 2000/1 were covered by the same volume, and that 2005 and 2006 went by without yearbooks largely because of my commitments towards my dying father.

However, the series was resumed last year, and it is with a modicum of pride that I can admit that this is the first issue since the debut that has actually appeared on time.

I am very pleased with the wide variety of articles in this volume, and whilst it is a very happy task to welcome back old yearbook stalwarts like Dr Karl Shuker and Richard Muirhead, it is an equally happy task to be able to introduce new writers such as Michael Woodley and Tony Lucas.

The keen eyed amongst you will notice that this year's volume is co-edited with my wife, rather than with Richard Freeman. This is for a number of reasons. Richard was trekking across the grasslands of Guyana while we put this book together, but far more importantly, Corinna can actually spell (which makes her almost unique within the annals of the CFZ).

This is a particularly fine edition, and I am already looking forward to compiling next year's book.

Jon Downes,
(Director, CFZ)
Woolfardisworthy,
North Devon
December 4th 2007

A TRIO OF MYSTERY CATS AT LONDON ZOO.

Dr Karl P. N. Shuker

London Zoo is famous for having housed, during its long and distinguished history, a number of spectacular animals that, sadly, are now extinct, such as the quagga, thylacine, bubal hartebeest, and Burchell's zebra. Less well-known, conversely, is that it has also been home to some creatures of appreciable cryptozoological interest, including the following trio of mystery cats.

During my extensive researches into cryptozoological felids, I have been the first person to publish a detailed modern-day account regarding any of these animals (and, indeed, the only person so far to have published anything at all concerning the third of them), but this is the first time that my writings on all three have been brought together within a single article, which also includes an illustration of each one.

THE WOOLLY CHEETAH – AN ENDURING ENIGMA

Following its original documentation by Philip Sclater on 19 June 1877 within the *Proceedings of the Zoological Society of London*, accompanied by a beautiful full-colour painting of the specimen in question, and a few subsequent late-19th-Century references to it, the woolly cheetah remained in almost total obscurity for over a century, until I

prepared what remains the most detailed account of this curious felid ever published, which duly appeared in my first book, *Mystery Cats of the World* (1989), and which I am now reprinting here. I have also documented it since then in lesser detail within various other publications of mine, including my eighth book, *Mysteries of Planet Earth* (1999), which republished the original colour painting for the first time since 1877.

In that year, Philip Sclater, secretary of the Zoological Society of London, recorded within its *Proceedings* the acquisition by London Zoo of a most unusual cat - male and apparently not fully grown which he described as follows:

> *It presents generally the appearance of a cheetah (Felis jubatus) [its old scientific name], but is thicker in the body, and has shorter and stouter limbs, and a much thicker tail. When adult it will probably be considerably larger than the Cheetah, and is larger even now than our three specimens of that animal. The fur is much more woolly and dense than in the Cheetah, as is particularly noticeable on the ears, mane and tail. The whole of the body is of a pale isabelline colour, rather paler on the belly and lower parts, but covered all over, including the belly, with roundish dark fulvous blotches. There are no traces of the black spots which are so conspicuous in all of the varieties of the Cheetah which I have seen, nor of the characteristic black line between the mouth and eye.*

Evidently this brown-blotched felid appeared very different from the usual form - to the extent that Sclater stated that it was impossible to associate it with this. Instead, he proposed for it the temporary name of *Felis lanea*, the woolly cheetah. It had been obtained from Beaufort West, South Africa, and, as Sclater himself remarked: "It is difficult to understand how such a distinct animal can have so long escaped the observations of naturalists".

One other matter is also difficult to understand, and remains a source of confusion concerning this mystery cat. Sclater referred to its markings as 'blotches', but in the painting that accompanied this report the creature was depicted with numerous tiny spots!

A year later, Sclater noted in the Society's *Proceedings* that he had received a letter from a Mr E.L. Layard, informing him that a second woolly cheetah was currently preserved in the South African Museum. As with the first, it had been procured from Beaufort West. It had been killed by Arthur V. Jackson who, like Layard himself, assumed that it was an erythristic (abnormally red) variant of the normal cheetah. At the end of this item, in answer to an enquiry by Layard, Sclater recorded that the claws of the London Zoo specimen were non-retractile.

Sharing Sclater's own bewilderment as to how so large and unusual an animal could have evaded scientific detection until then, many zoologists had grave reservations concerning his optimism that the woolly cheetah constituted a totally separate species. In 1881, St George J. Mivart commented that the noted zoologist Daniel Elliot regarded this felid simply as a variety of the known cheetah species (curiously, Mivart added a stripe to one side of the woolly cheetah's muzzle in his book *The Cat*, not mentioned by Sclater and in any event highly abnormal, thereby confusing the issue even further). By then, London Zoo's specimen had died, and Elliot's opinion received support from the discovery by Oldfield Thomas of the British Museum (Natural History) that this cat's skull did not differ from that of any other cheetah.

In 1884 Sclater recorded a woolly cheetah skin sent to him by the Reverend G. Fisk, again obtained from Beaufort West. In comparison with the zoo specimen, this example was more distinctly spotted, less densely furred and rather smaller in size. Reverend Fisk believed that these differences were due to the specimen being a female, an explanation accepted by Sclater, who felt that this new skin consolidated his opinion concerning the woolly cheetah's separate status. The rest of the scientific world, conversely, remained unconvinced, so that since then it has been regarded as nothing more than an unusual variant of the typical cheetah species.

According to mammalogists Daphne Hills and Dr Reay Smithers, this odd form no longer occurs in Beaufort West. Presumably, therefore, it is extinct, and the chance to investigate further its precise taxonomic status similarly lost. Or is it? The British Museum owns the skin of the zoo specimen - perhaps it may be possible to carry out genetic tests upon small samples of this.

The woolly cheetah might indeed be nothing more surprising than an atypical colour morph – perhaps a partial albino as suggested by king cheetah researcher Lena Bottriell and by cat geneticist Roy Robinson among others, or an erythristic mutant as assumed by Layard and by Jackson. At the same time, Sclater's more radical views can also be appreciated, because this cat form differs from the typical cheetah not only in colour and markings but also in fur density and even in relative limb length. Simple colour variants do not generally exhibit such pronounced differences as these from normal individuals of the same species. Its shorter limbs suggest a non-cursorial life - could the woolly cheetah possibly have been a forest form? It is worth noting that a 'lion-like forest cheetah' known as the kitanga was described during the 20th Century's early years to Major G. St John Orde-Brown by the Embu natives of south-eastern Kenya, and that a comparable felid has occasionally been reported from the little-explored forests of Senegal, West Africa, where the typical cheetah *A. j. hecki* is extremely rare.

THREE CATS IN ONE – LONDON'S AMAZING LI-JAGUPARD!

Over the years, a vast array of interspecific feline crossbreeds have been produced in captivity, but none, surely, can boast not only the complicated genetic identity but also the exotic, much-muddled history of this second London Zoo crypto-cat. For a time, it thoroughly perplexed everyone who saw it, but following the remarkable denouement concerning its true zoological status this extraordinary animal sank into relative obscurity. In 1968 it resurfaced briefly, when Mainz University zoologist Dr Helmut Hemmer published a short article concerning it in the German-language journal *Säugetierkundliche Mitteilungen*, but afterwards it attracted little if any further attention until I prepared an account of it that appeared in a *Wild About Animals* magazine article of mine from January 1996 regarding leopons (lion x leopard hybrids), and is now reprinted here in slightly expanded form. It also features in my new book on mys-

terious and mythological felids, entitled *I Thought I Saw the Strangest Cat*, due for publication by the CFZ Press in the not-too-distant future.

One of the most remarkable episodes appertaining to lion x leopard crossbreeding began almost exactly a century ago – with the arrival for identification purposes at London Zoo in 1908 of a singularly mysterious felid that resembled a slim, gracile lioness but was elegantly dappled with large brown rosettes recalling those of the Himalayan snow leopard! Its owner, a well-known London-based animal dealer called Hamlyn, who in turn had purchased it from an African seller, asserted that the seller had claimed it had been captured in the junglelands of the French Congo (now the People's Republic of Congo) and represented a wholly new species. This dramatic allegation naturally attracted a great deal of media publicity and scientific speculation, and was also why Hamlyn had brought the cat to the zoo, in order for it to be formally examined and its zoological identity conclusively ascertained.

However, London Zoo's superintendent, cat expert Reginald Pocock, was sceptical of such a grandiose claim from the original African seller, dismissing this feline enigma in letters to *The Field* as a leopon.

After being on display at the zoo for just a fortnight it was sold by Hamlyn at auction (for the incredible sum of 1030 guineas due to its high media profile), and after being shown at the White City it was removed from London to Glasgow. Tragically, this intriguing animal died only a few years later – reputedly killed by a lion that broke through into its cage from a neighbouring enclosure (yet, oddly, there was no sign of injury to its pelt, which was duly preserved) – and was later exhibited as a mounted taxiderm specimen at France's National Museum of Natural History in Paris.

By then, however, the true history of this controversial cat had been uncovered, and its extraordinary identity at last exposed and documented by Pocock in *The Field*. At the turn of the 19th Century, a male jaguar had mated with an Indian leopardess at Chicago's Lincoln Park Zoo, the result of which was a litter of three 'jagupards' - one male and two females.

All three were later sold to a travelling menagerie, but whereas the male was killed by a lion, his two sisters grew to adulthood, and both of them mated with a lion. Remarkably, these matings were viable (most first-generation big cat hybrids are sterile), yielding several cubs – and it was one of these that found its way to London Zoo, deceitfully labelled by this much-crossbred cat's original seller as a new species.

Bearing in mind that this amazing animal was the complex product of genetic intermingling between three different species of big cat – lion, jaguar, and leopard – it is little wonder that it seemed so exotic in appearance, and engendered such confusion. After all, it isn't every day that a li-jagupard (also dubbed a lijagulep) turns up in London – or anywhere else, for that matter!

-CFZ YEARBOOK 2008-

LONDON ZOO'S LONG-FORGOTTEN BLACK PUMA – HOW I REDISCOVERED IT AT HAY-ON-WYE!

The third of London Zoo's mystery cats to be documented here is, for me, the most exciting – because I personally discovered it, or, to be precise, rediscovered a spectacular piece of evidence confirming its erstwhile existence after it had apparently been totally forgotten for over a century. Certainly, to the best of my knowledge this exceptional, highly cryptozoologically-significant creature had never once been so much as mentioned by anyone, anywhere, in all that time until I exclusively documented it in an extensive *Beyond* magazine article of mine surveying a very diverse range of aberrantly-coloured mystery cats and published in October 2007. The relevant section from that article is as follows.

The two most commonly-voiced identities for Britain's elusive ebony-furred mystery cats are escapee black panthers (i.e. melanistic leopards) and escapee black (melanistic) pumas. Yet whereas the former is plausible, the latter is little short of impossible, for one extremely good, fundamental reason. Despite the fact that the puma has the greatest native distribution range of any modern-day species of wild cat, recorded from the northernmost regions of North America to the southern tip of South America, the number of confirmed black pumas can be counted on the claws of one paw! No preserved

specimen exists, and only one clear photo is known – of a dead specimen shot in Costa Rica during 1959, which has been published in Jim Bob Tinsley's comprehensive book *The Puma: Legendary Lion of the Americas* (1987), and also in a *BBC Wildlife* article from 1995 by the late J. Richard Greenwell.

Moreover, whereas most melanistic cat forms are uniformly black all over, so-called black pumas only have black upperparts; their underparts are noticeably paler, usually slate-grey or dirty cream. This, therefore, is another reason for discounting such cats as the identity of Britain's pantheresque mystery cats - which, just like black panthers, are black all over.

Yet although exceedingly scarce today, black pumas do seem to have been more common in past ages, certainly in South (even if not in North) America, because there are a number of reports and even one or two early illustrations of such cats, sometimes dubbed 'couguars noires', in archaic natural history tomes. And these reports and illustrations often compare closely with the few verified modern-day specimens (although in some cases there appears to have been confusion between, and amalgamation of, black puma and black jaguar reports, as in the case of the jaguarete, documented by me in *Mystery Cats of the World*).

As far as I was aware, however, no confirmed black puma had ever been kept in captivity, at least not in Europe. But all that changed a while ago during one of my numerous visits to one of my all-time favourite places – Hay-On-Wye, Herefordshire's world-famous 'Town of Books', nestling on the Welsh border.

In addition to over 40 bookshops, this small town also has shops devoted to antiquarian prints. As an avid collector of such items, I was browsing in one of these shops when I came upon a truly remarkable example – remarkable because it is not often that an antiquarian print depicts a cryptozoological cat! The print in question, which was an original hand-coloured copper engraving dating from 1862, and which I naturally lost no time in purchasing, is duly reproduced here (its earlier appearance, in my *Beyond* article, where it was published in its original full-colour format, may well have been the first time that it had ever been published anywhere), and appears to portray a bona fide black puma.

Certainly, it comes complete with jet-black upper parts, slaty-grey under parts, and white chest – very different from normal pumas, which are either tawny brown or silver-grey (the puma exhibits two distinct colour morphs), but matching precisely those few confirmed black puma specimens. Most interesting of all, however, is the engraving's caption: "The Puma. In the Gardens of the Zoological Society". This means that if the puma in the engraving has been coloured accurately, and there is no reason why it should not have been, a black puma, that most mysterious of mystery cats, was once actually on display at London Zoo!

I am currently pursuing this line of enquiry further (although whether sufficiently-detailed records for that period still exist is doubtful). Who knows, this astonishing black puma may even have been in the enclosure next door to the zoo's unique captive woolly cheetah!

Indeed, all that remains to be said here is that it would clearly be very worthwhile not only pursuing the history of London Zoo's black puma (as I am now doing) but also investigating whether, in its long, unparalleled history as a repository for diverse creatures from all corners of our planet, this most significant of world zoos may at one time or another have exhibited other mysterious but presently forgotten creatures deserving of attention from modern cryptozoology.

REFERENCES

ANON. (1908). Supposed hybrid lion and leopard. *The Field*, vol. 111 (No. 2887; 25 April), p. 711.
GREENWELL, J. Richard (1995). The puma plot thickens. *BBC Wildlife*, vol. 13 (No. 5; May), p. 26.
HEMMER, Helmut (1968). Mitteilung über einen Bastard Löwe x (Jaguar x ? Leopard) – *Panthera leo x (P. onca x ? P. pardus)*. *Säugetierkundliche Mitteilungen*, vol. 16, pp. 179-182.
LYDEKKER Richard (1893-1894). *The Royal Natural History*. Frederick Warne & Co (London).
LYDEKKER, Richard (1895). *Cats – Allen's Naturalist's Library*. J.F. Shaw (London).
MIVART, St George J. (1881). *The Cat*. John Murray (London).
SCLATER, Philip L. (1877). [Description of the woolly cheetah.] *Proceedings of the Zoological Society of London*, 19 June, pp. 532-533.
SCLATER, Philip L. (1878). Mr. P.L. Sclater on Felis lanea. Ibid., 18 June, pp. 655-656.
SCLATER, Philip L. (1884). [The woolly cheetah.] Ibid., 4 November, p. 476.
SHUKER, Karl P.N. (1989). *Mystery Cats of the World: From Blue Tigers To Exmoor Beasts*. Robert Hale: London.
SHUKER, Karl P.N. (1996). Leopons a-leaping. *Wild About Animals*, vol. 8 (No. 1; January), pp. 50-51.
SHUKER, Karl P.N. (1999). *Mysteries of Planet Earth: An Encyclopedia of the Inexplicable*. Carlton Books: London.
SHUKER, Karl P.N. (2007). Technicolor tigers and other multicoloured mystery cats. *Beyond*, No. 9 (October): 56-63.
TINSLEY, Jim Bob (1987). *The Puma: Legendary Lion of the Americas*. Texas Western Press: El Paso.

THE STRANGE ANIMALS OF SWANSEA BAY
Oll Lewis

Dylan Thomas called Swansea "The death-place of ambition". This might have been a bit over-critical of the city whose main claim to fame was Swansea-Jack, the only dog to be awarded the canine Victoria Cross on two separate occasions for risking his life to save people from the waters of Swansea bay. As a reminder of Jack's short life of heroism, (he died when he was 7 years old after eating rat poison) Swansea residents are known as "Jacks" to this day. Jack's feats of bravery were considered unusual in his day but there are, and have been, much more unusual animals reported in the area of Swansea Bay.

One particular strange animal hotspot mentioned several times in folklore was the spot where four waterfalls met near Glyn Neath. The waterfalls of the rivers Perddyn, Little Neath, Mellte and Hepste can lay claim to having been the location for sightings of a both a gwiber and a water horse.

The water horse, or ceffyl-dwr in Welsh, that was said to have based itself near the waterfalls has several mentions in local folklore, the most well known tale having appeared in Marie Trevelyan's "Folk-lore and folk-stories of Wales" and is said to have taken place in the first half of the 19th century.

A traveller on a long journey became weary and decided to rest in a shady spot near the waterfalls. During his resting he saw a horse coming up from behind a cascade. The bright white animal shook the water from its mane and sauntered up the slope to where the man lay. The weary traveller was tempted by the fine stallion he saw before him and after checking that there appeared to be no-one else in the vicinity he mounted the creature, which seemed too docile to be wild and used its silky soft mane in lieu of reins. From the traveller's point of view this seemed too good to be true, a well-behaved, well-kept and powerful horse had come his way just as he had needed it. Then the traveller noticed that the animal was moving rather fast, which the traveller thought was rather good as he would reach his destination much quicker. The traveller's good spirits soon turned to fear as the horse sped up even more, travelling at a lightning pace for miles and miles. By the time the horse finally slowed down enough for the traveller to jump off without breaking his neck the horse had been running for an hour and reached the village of Llanddewi Brefi in Cardiganshire, two counties away! When the man recovered he was shocked to see the horse slowly dissipate into a mist.

As well as a gwiber being said to have made its home in the area of the four waterfalls gwibers were also said to have been found on the Gower peninsular. Gwibers are winged serpents that were said to vary in size from about 15 centimetres (half a foot) long to monstrous sizes of ten metres. Small gwibers, like the gwibers of Penllyne (mentioned in my 2007 CFZ yearbook article, The Mystery Menagerie of Glamorgan) were said to be of no direct threat to man, but a serious farm pest, larger gwibers were said to be very dangerous creatures indeed. None of the gwiber sizes in the areas were recorded into modern times but, because there was a legend that the gwiber from the waterfalls guarded a treasure in the caves and gorges of the vicinity, it was probably reputed to have been fairly large and formidable.

Sadly the nature of folklore means that evidence of the existence of the preceding animals is scant at best; however, Swansea and the surrounding area can lay claim to being the home of an unusual animal that only the most blinkered sceptic could doubt the existence of.

Few people will have heard of the town of Baglan and an even smaller number know that a big cat is said to roam the district. The few mentions the cat has had in the press outside of the local newspapers often contradict each other so fewer people still know the true importance of the Baglan big cat.

In April 2005 a teenager named Nathan Flavell was working on English homework with his tutor, Janice Lewis, when he spotted something through his kitchen window.

Nathan described the creature he saw on the mountain side as being about 3 feet tall and jet black, the sun was behind the creature so it is unclear whether the jet black colouration was the creature's natural colour or weather it just appeared to be as a result of the sun silhouetting the animal. It had a muscular face, short black ears and a thick chunky tail. The sighting was made from less than 50 yards away.

Nathan was adamant that the animal was not a dog as a friend of his has a rottweiler and the animal was larger and of a different body shape. The animal left after Nathan had been staring at the animal for around 15 minutes, but not before his tutor also caught a glimpse after noticing Nathan was distracted by something.

A waterfall on the Little Neath river where a water-horse appeared to a weary traveller

The Swansea Jack memorial

Swansea Jack; Britain's most courageous canid

A painting of Swansea Jack with some puppies he saved from drowning in the dock

A view of Swansea Bay

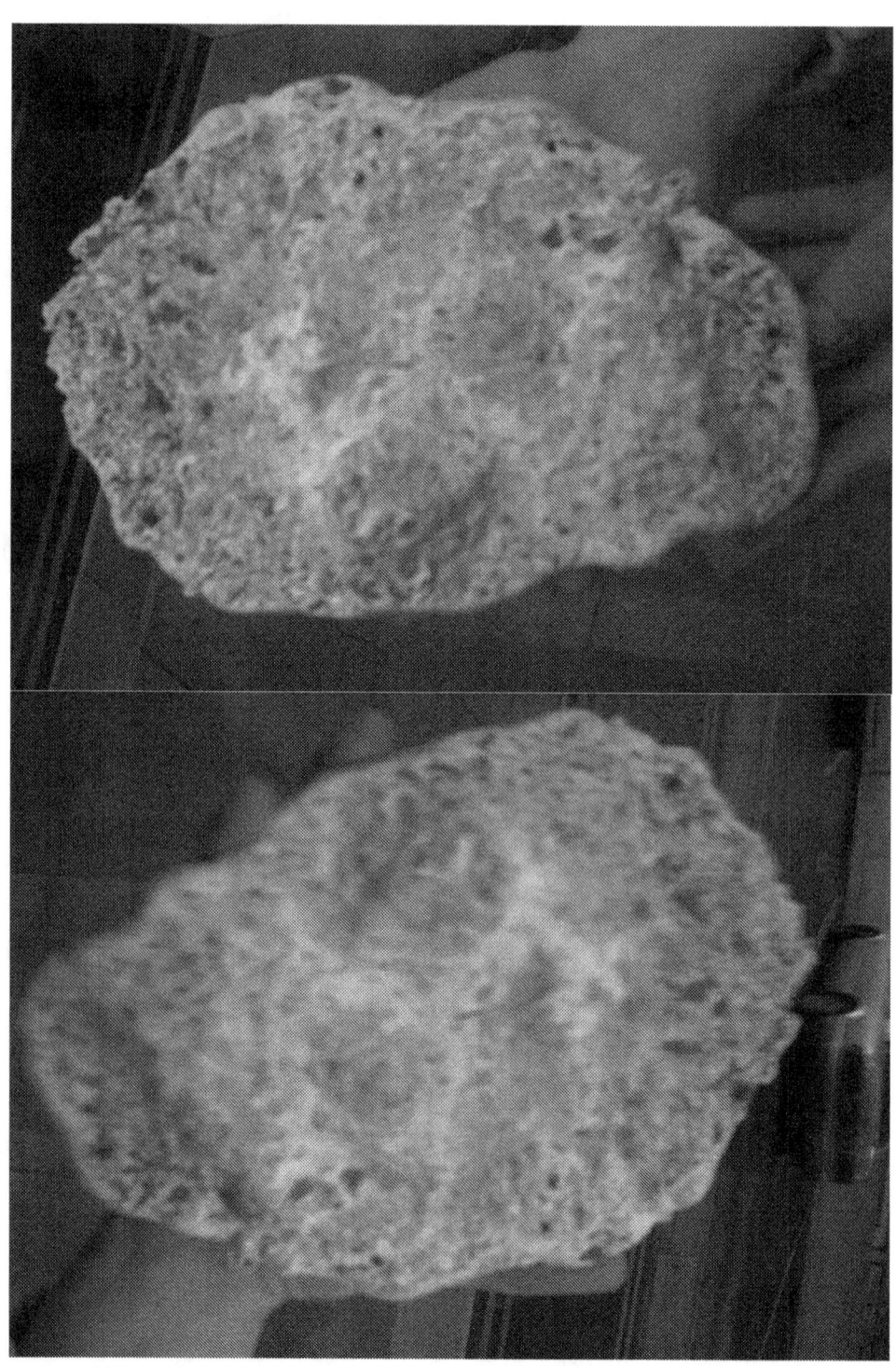

Plastercast made from prints found in Baglan by Neath-Portalbot police, Photographed by Oll Lewis.

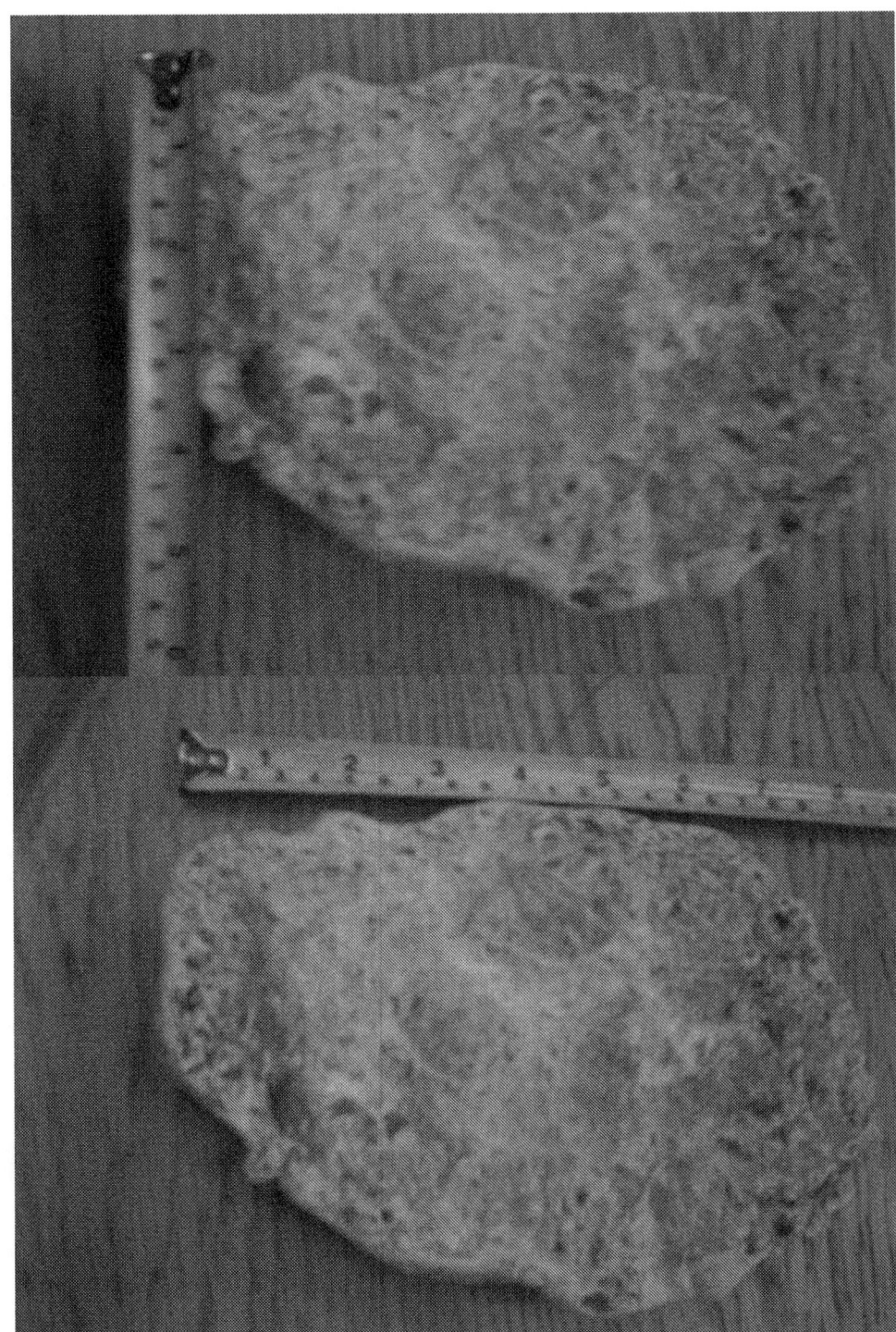

"Less than a month later the superintendent of Neath Port Talbot police station himself came face to face with the beast, while out for an evening jog. When he came on duty the next day he took several officers to the spot to photograph the site and search for more evidence of the animal. While there, officers found several tracks, which they photographed, and one print made deep in soft ground at the side of the path which they took a plaster cast of. This plaster cast also contained hair from the animal. When their examination of the scene was complete the police were in little doubt that there was a big cat roaming the area and reported it to the Welsh Assembly Government (WAG) wildlife department, which holds jurisdiction over these matters in the principality as DEFRA does in England."

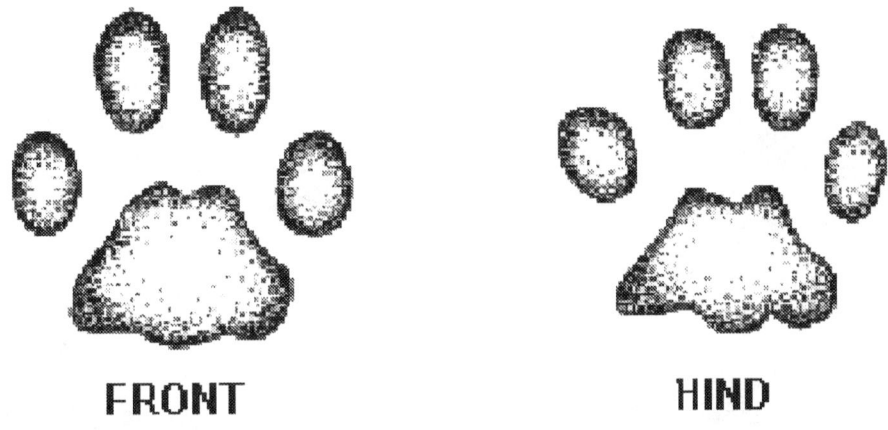

Drawing of bobcat paw prints (above), bobcat prints in-situ, note that hind paw print has been made on top of the front (below).

American bobcats in areas of human habitation

She told the local paper that she had seen the creature clearly for around 10 seconds and that it "had that big cat thing about it". As well as having that big cat thing about it as she had put it, the thing that most stuck in her mind about the animal was its black pointy tipped ears.

Nathan's mother, Susan Flavell, the deputy headmistress of a local comprehensive school, called the police when she was told of the sighting as many people use the area for walking their dogs and a detective from Neath Port Talbot CID attended the scene interviewing the boy and his tutor.

The police probably didn't realize at the time just how involved in the case they would become.

Things went fairly quiet for some months and many in the town probably either thought that the big cat that the papers had said was a puma had moved on or had been a misidentification. But early in October a local man near to where the Flavell sighting had been made, made a gruesome discovery. One of his pet goats had been attacked and partially eaten. Not making the connection with the big cat sighting at the time the man buried the carcass. He informed the police a few weeks after when he had realised the sighting had been made very close to where he had found the goat's body.

Police exhumed the body in the hope that they could confirm the reports the owner had given them about the state of the body, establish the cause of death and confirm whether or not a big cat had been involved, either in the animal's demise or by eating the flesh after death. The police also put out a press release appealing for witnesses on the 25th of October. Unfortunately the goat was in too advanced a state of decomposition to be able to make any conclusions whatsoever, so it was reburied.

Less than a month later the superintendent of Neath Port Talbot police station himself came face to face with the beast, while out for an evening jog. When he came on duty the next day he took several officers to the spot to photograph the site and search for more evidence of the animal. While there, officers found several tracks, which they photographed, and one print made deep in soft ground at the side of the path which they took a plaster cast of. This plaster cast also contained hair from the animal. When their examination of the scene was complete the police were in little doubt that there was a big cat roaming the area and reported it to the Welsh Assembly Government (WAG) wildlife department, which holds jurisdiction over these matters in the principality as DEFRA does in England.

I later obtained a copy of the WAG reports and was able to examine the scene of incident photos and plaster cast under a freedom of information request and from the documents it seems clear that the WAG took a more sceptical line to the police in their investigation.

The Superintendent's sighting was recorded as follows in the WAG report:

The Superintendent was jogging with his Jack Russell terrier pup on a footpath in the wood when he saw a large animal cross the path. The animal jumped up from the slope below, landed on the path and with one leap jumped up onto the slope above and disappeared. The Superintendent was convinced that it was a large cat. Police Scenes of Crime returned the scene the following morning and searched the area and took a plaster-cast of a large paw print and several photos of prints. It is also believed that they have samples of hair from the scene. Plaster cast was described as being bigger than a fist. It was arranged to meet Sgt Ian Guildford at Neath PS on 23 November. In the meantime, Neath Police have issued a Press Statement and the sighting was mentioned on the evening news. 23 November - Superintendent was adamant that he had seen a big cat. The animal was brown with a long neck - he had no recollection of seeing a tail. Animal was seen for 3-5 seconds. He was jogging down to the entrance when the animal appeared. Before this he was aware of other people in the wood below the track who he thought were exercising/working their dogs, possibly spaniels.

Plaster-cast was shown - appeared to be one print superimposed on another and not very clear. Similarly with the photos - not very clear and difficult to distinguish obvious prints. Site was visited - the woodland backs on to a housing estate and is close to the incident where it was alleged that a pet goat had been killed by a big cat.

The report then detail's the WAG wildlife departments own investigation of the path several days later with Sgt Ian Gilford of the Neath Port Talbot police several days after the sighting and their failure to find additional prints in the area.

The conclusion of the WAG report reads:

"Unsure as to what was sighted. Very unlikely that any big cat would venture so close to human habitation, unless it was preying on domestic animals. No reports of missing dogs/cats etc in the area. Police will keep a close eye on any reports of sightings in the area, for the next 2 weeks."

It is not unlikely that all big cat species would not venture close to human habitation. Some species commonly hunt or scavenge in farmland and the urban fringe, for example Bob-cats in America. Another of the conclusions of the report, that no cats or dogs are known to be missing in the area being somehow evidence against a big cat in the area is nonsensical; a big cat is more likely to go after easier to catch and plentiful prey such as wild rabbits which are lower down the food chain than a cat or dog that could fight back.

It is possible the report was an attempt to pour cold water on the incident in-case a panic ensued locally, it is certain that newspapers were also briefed that the WAG had investigated the matter and found "no evidence". While it was true that the WAG's later investigation had found no evidence this was not the case in the local police investigation.

The plaster cast is very large, 2 ½ inches across, much bigger than that of a large dog, and the animal's paw had sunk into the soft mud right to the top of the foot, also giving an impression of the front of the toes where it exhibits retracted claws. The cast is shaped like a cat's paw print and the print of a smaller hind paw can be seen superimposed inside the larger print which is further evidence that it was made by a big cat because cats hind paws step in the tracks of the forepaws as they walk and dogs do not.

From the Superintendent's description and the size and shape of the tracks it is likely the animal seen was a bob-cat (*Lynx rufus*). This is further supported by the results of DNA tests undertaken on the hairs found in the cast. According to a statement by acting inspector Hugh Griffiths of Neath Port Talbot police the conclusion reached by experts from the University of Wales was that the hair was from a lynx or mountain lion.

Such is the weight of evidence behind it, the Baglan Bobcat is one of the best examples of a big cat in Britain, and one of the most unusual animal inhabitants of south Wales.

PLANET OF THE APE MEN

Nick Redfern

Born on September 21, 1866 in the English town of Bromley, novelist H. G. Wells was the son of a professional cricket-player and a housekeeper, and made an untold number of contributions to the world of science-fiction that have, arguably, never been eclipsed. In 1883, and after having first been employed as a draper's assistant (an experience that he detailed in his 1905 book, *Kipps*), Wells became a teacher-pupil at Midhurst Grammar School, and subsequently obtained a scholarship to the Normal School of Science in London, where he studied biology. His interest faltered, however, and in 1887 he left without obtaining his degree.

Wells then taught in private schools for four years, not taking his B.S. degree until 1890. In the following year, Wells settled in London, married his cousin Isabel, and continued his career as a teacher in a correspondence college. He would, however, later leave Isabel for one of his students, Amy Catherine, whom he married in 1895. But the real turning point for Wells had come in 1893, when he embarked upon a career as a full-time writer.

The author of such classics as *The Time Machine, War of the Worlds, The Invisible Man*, and *The First Men on the Moon*, Wells was a true prophet, anticipating both the advantages and disadvantages that the human race would face as technology and science began to increase at an exponential rate. Indeed, *The Time Machine* was actually a less-than-subtle parody of British class-divisions; while his lesser-known title of 1908, *The War in the Air*, revealed Wells' wholly justified concerns that the development of aircraft would inevitably lead to their use as devastating tools of warfare. It was only to be a few short years before the carnage of the First World War began, and Wells' fears

were proven to be utterly correct.

In the aftermath of the War, Wells turned his hand to non-fiction, writing such titles as *The Outline of History*, *The Science of Life*, and *Experiment In Autobiography*. But Wells' fears with respect to what the future might bring never really left him: in 1939 he penned *The Holy Terror* which was a study of the psychological development of a modern dictator based on the careers of Stalin, Mussolini, and Hitler.

Wells lived throughout the Second World War in his house in Regent's Park, London - refusing to be intimidated by Hitler's minions. Wells' final book, *Mind At the End of its*

Tether was published in 1945 and expressed extreme pessimism with respect to humankind's future prospects – or lack of them. He died in London in the following year, on August 13.

But perhaps most intriguing – and certainly most relevant to this edition of the *CFZ Yearbook*, as will shortly become graphically apparent – was Wells' 1896 sci-fi novel

The Island of Dr. Moreau. As was the case with so many of his other works, *The Island of Dr. Moreau* related some of Wells' ever-present worries about the ways in which science was advancing.

The book is presented in the form of a discovered manuscript that is introduced by the nephew of the narrator, and that is then transcribed for the benefit of the reader. Following a catastrophic shipwreck, Edward Prendick finds himself marooned on an unknown island where, to his horror, he learns of the diabolical 'vivisections' of Dr. Moreau of the title of the book. In an attempt to create a race free of malice, Moreau explains to the shocked Prendick that he has achieved what had up until then had been presumed completely impossible: the transforming of wild animals into strange creatures that are 'human in shape, and yet human beings with the strangest air about them of some familiar animal'.

Prendick slowly begins to remember the bizarre exploits of the infamous Moreau from more than a decade earlier, when he was known for his rumored fascination with the world of experimental, fringe science.

Moreover, Prendick recalls in the book how a journalist who had published a pamphlet called *The Moreau Horrors!* subsequently exposed Moreau, who was then shunned by the scientific community, fled England, and was never heard from again. Until now, that is.

Since Prendick is stranded on the island, Moreau and his assistant, Montgomery, eventually decide to confide in him the full and shocking story of their eleven years of work on the island that involves the man-beasts being held in check by a series of prohibitions that have been 'woven into the texture of their minds beyond any possibility of disobedience or dispute'.

The laws are: 'Not to go on All-Fours'; 'Not to Suck up Drink'; 'Not to eat Fish or Flesh'; 'Not to claw the Bark of Trees'; and 'Not to chase other Men.' Always, the creatures are required to endlessly repeat: 'That is the Law. Are we not men? That is the Law. Are we not men? That is the Law. Are we not men?'

Not content with playing God, Moreau actually tries to turn himself into a living deity by having the man-beasts worship no-one, and nothing, aside from him. Unfortunately, and to his ultimate cost, Moreau learns that unless he continues to mold, exploit and reign in the natural animal-instincts of his diabolical creations on a regular basis, they begin to revert to their bestial ways.

As a result, a disturbing undercurrent of revolution begins to stir within the strange world of the 'manimals' and Moreau and Montgomery are violently slain by their creations.

Prendick is spared, however, and lives for a while with the new rulers of the island until, unable to relate to them after their all-but-complete reversal to animalistic states, he finds an abandoned boat and succeeds in leaving the confines of the island. Prendick is shortly thereafter rescued; however, his strange tale is assumed to be nothing more than the ravings of a madman driven insane by a near-year-long period of isolation on the island.

As a result the dark secrets of Dr. Moreau remain precisely that – secrets.

At the time that Wells wrote *The Island of Dr. Moreau*, England's scientific community was engulfed by debates on the controversial issue of animal vivisection, and it was largely this issue that led Wells to write his novel.

And only two years after the book was published – that graphically highlighted in its pages such issues as vivisection, irresponsible and unregulated scientific research, religion, and eugenics - the British Union for the Abolition of Vivisection was formed and became a noted and forceful official body; which suggests that perhaps someone recognised that the scenario that Wells portrayed might one day become a reality.

A somewhat similar scenario was portrayed in Pierre Boulle's classic 1963 sci-fi story *Planet of the Apes* that was turned into a Hollywood blockbuster film five years later starring Charlton Heston. In the Oscar-winning film version, Heston's character, Taylor, is an astronaut who travels across the depths of space to a planet where apes are the dominating life form and man is little more than a savage creature that acts as sport, slave, and medical guinea pig for the ape civilization. Only at the end of the film does Taylor discover the shocking truth: that he has not crossed space at all. Rather, he is back on Earth, an unknown number of years in the future, after a devastating nuclear war has obliterated practically every memory and remnant of human civilization – aside from a solitary, and partly-destroyed, Statue of Liberty that Taylor finds half-buried in the sands of a new landscape carved by the destructive effects of the war.

With the success of the first *Apes* movie, four more followed, as did two television spin-offs, and a *Marvel* comic book. The developing story that is revealed demonstrates that after a catastrophic plague wipes out all of the cats and dogs on the planet, human beings begin to take primates first as pets for company, then as slaves who they exploit as a result of their own laziness. The physical appearance of the apes begins to change over time, and they ultimately develop the power of speech. Inevitably, the apes, realising their newly found potential, revolt against their human masters. And with nuclear war looming, the countdown to the end of human civilization, and the birth of a whole new ape-controlled society, begins.

But surely the collective scenario presented in *The Island of Dr. Moreau* and in *Planet of the Apes*, of wild animals becoming more human that bestial, could only ever be considered sci-fi? Probably; but not everyone thought so.

Secret government files generated in Russia in 1926 under the regime of Premier Josef Stalin reveal the details of an astonishing and shocking story that eerily parallels the scenario presented in Wells' *The Island of Dr. Moreau* and *Planet of the Apes*. According to the formerly classified records, Stalin had a crazed idea to try and create an army of creatures that would be a combination of half-ape and half-man, and that would be utterly unbeatable on the battlefield.

As a result of Stalin's plans, Ilya Ivanov, the former Soviet Union's top animal-breeding expert at the time, was personally told by Stalin: 'I want a new invincible human being, insensitive to pain, resistant and indifferent about the quality of food they eat.' Somewhat shrewdly, and per-

Josef Vissarionovich Dzhugashvili (Georgian: *Ioseb Besarionis Dze Jughashvili*; Russian: Иосиф Виссарио́нович Джугашви́ли, *Iosif Vissarionovich Dzhugashvili*) (1878–1953), better known by his adopted name, Joseph Stalin, was General Secretary of the Communist Party of the Soviet Union's Central Committee from 1922 until his death in 1953. Although Stalin's formal position originally had little significant influence, his office being nominally but one of several Central Committee Secretariats, Stalin's increasing control of the Party from 1928 onwards led to his becoming the *de facto* party leader and the dictator of his country, in full control of the Soviet Union and its people. **From Wikipedia, the free encyclopedia**

haps anticipating a scenario similar to the catastrophic ending in *The Island of Dr. Moreau*, Stalin added that the creatures should possess 'immense strength but with an underdeveloped brain'.

Certainly, in the eyes of Stalin if anyone could make the crackpot project succeed it was Ivanov. A highly regarded figure, he had established his reputation under the Tsar when, in 1901, he established the world's first centre for the artificial insemination of racehorses. But more important to Stalin was the fact that Ivanov had already tried to create a 'super-horse' by attempting to crossbreed such animals with zebras.

Despite the fact that the attempts to crossbreed a horse with a zebra failed completely, Moscow's Politburo forwarded Stalin's request to the Academy of Science with the order to build a 'living war machine'; an order that came at a time when the Soviet Union was embarking upon a crusade to turn the world upside down, with social engineering seen as a partner to industrialisation.

In addition, Soviet authorities were struggling at the time to rebuild the Red Army after the devastation of the First World War, and there was also intense pressure to find a new labour force, and particularly one that would not complain. As a result, in the warped mind of Josef Stalin, the secret creation of a super race of hybrid creatures that combined the intelligence of human beings with the physical strength of some of the larger primates, such as gorillas and chimpanzees, seemed to be the perfect antidote to every problem.

The Russian scientific community swung into action and Ivanov was quickly dispatched, with $200,000 in his pocket no less, to West Africa where the first such experiment was planned: namely the impregnation of a number of chimpanzees with human sperm. Ivanov's now-archived reports reveal that the Pasteur Institute in Paris, France secretly granted him permission to use their research station in Guinea, West Africa, for ape-breeding research.

Ivanov advised the Politburo that: '...the biggest problem is to catch living females.' As a result, Ivanov's team learned that the answer to this tricky problem was to burn the trees and chase the apes into cages as they scampered down the trunks. Ivanov also reported, somewhat disturbingly, on the fact that his team had 'seized' a number of local African women in the area who were 'to be impregnated with ape sperm'. No pregnancies resulted. More ambitious plans to impregnate female gorillas with human sperm also ended in complete failure.

At the same time, a centre for such experimentation in Russia was stealthily established in Stalin's birthplace of Georgia, where the super-apes were to be raised if impregnation was ever seen to be successful. Unsurprisingly, none of the West African experiments succeeded. Undaunted, however, Stalin pressed on with an even more controversial plan: he arranged for a number of women 'volunteers' in Russia to be impregnated with monkey sperm in an effort to determine whether or not following this particular route would prove to be more successful. Again, it was not. That such experimentation did proceed, however, is not in doubt: only recently, workmen engaged in the building of a children's playground in the Georgian Black Sea town of Suchumi found a plethora of ape-skeletons, and an old abandoned laboratory.

In the eyes of the ruthless Stalin, and a result of the resounding failure to create an army of man-beasts, Ivanov was now in complete disgrace. As a result, Stalin sentenced Ivanov to five years in jail, which was later commuted to five years' exile in the Central Asian republic of Kazakhstan in 1931. He died a year later, after falling sick while standing on a freezing railway platform. And while the whole, strange and secret affair of Russia's super-apes can today be viewed as farcical from beginning to end, there is a serious point to all of this, and it was one that H.G. Wells recognised more than a century ago: that ruthless men without morals, operating in complete secrecy, and with access to advanced science and technology, might be the undoing of our whole society. We would do well to remember the concerns of this sci-fi visionary, if only to ensure that one-day another Stalin does not try something similar – and succeed.

And there is a very intriguing footnote to this story: H. G. Wells visited the Soviet Union in 1934, and, on July 23rd of that year, personally interviewed Joseph Stalin. The conversation, lasting from 4.00 p.m. to 6:50 p.m., was recorded by Constantine Oumansky, then head of Russia's Press Bureau of the Commissariat of Foreign Affairs. As far as can be determined at least, the two men did not drift into a discussion of genetically created ape-men (their exchange was largely focused upon the then current state of the world from an economic perspective). However, perhaps Stalin – a voracious reader who is known to have read Wells' output – mused upon the possibility of informing Wells that the scenario the sci-fi author had presented within the pages of *The Island of Dr. Moreau* was something that he, Stalin, was secretly quite familiar with. Doubtless Wells would have considered this prime evidence that his worst fears and nightmares about science really were on the verge of coming all-too-horrifically true.

References:

1. www.online-literature.com/wellshg/
2. www.kirjasto.sci.fi/hgwells.htm
3. *Stalin's Army of Mutant Ape-Men*: http://news.sky.com/skynews/article/0,,30000-13482528,00.html
4. *Stalin's Mutant Ape Army*: www.freerepublic.com/focus/f-news/1543814/posts
5. *Joseph Stalin's and H.G. Wells, Marxism Vs Liberalism*, http://rationalrevolution.net/
6. *The Island of Dr. Moreau*, H.G. Wells, www.bartleby.com/1001/
7. *The Island of Dr. Moreau*, http://en.wikipedia.org/wiki/The_Island_of_Dr._Moreau

Nick Redfern is the author of many books on various aspects of Forteana, including Man-Monkey: In Search of the British Bigfoot (CFZ Press, 2007). He runs the American Office of the Centre for Fortean Zoology from his home in Dallas, Texas. Nick can be contacted at his website, www.nickredfern.com or at his cryptozoology blog, There's Something in the Woods: http://monsterusa.blogspot.com

TOWARDS A POSSIBLE CAUDATA IDENTITY FOR THE MONGOLIAN DEATH WORM:
Introducing the 'plausibility method' for identity theory formation amongst lesser known cryptids.
Michael A. Woodley

Introduction.

Few cryptids are as fearsome in reputation as this purported denizen of the Gobi desert. Described as being capable of discharging a lethal, acidic fluid and reputedly possessing a deadly electrical attack, the Mongolian death worm is allegedly capable of killing a person at a great distance with near instantaneous efficiency.

Despite the notoriety of this cryptid, very few theories have been posited as to its potential identity. This is something that the death worm has in common with many lesser known cryptids for which there exists very little data and correspondingly little idea as to how to go about researching them.

To address this issue, this monograph aims to introduce a new targeted research methodology in the form of the 'plausibility method', with the intention of a) suggesting a new salamander identity theory for the death worm, and b) showing how the incorporation of extrinsic factors into an identity theory, such as inferred evolutionary history and ecology, can allow for new research targets to be suggested.

Describing the indescribable.

The death worm has been described as being between two to five feet in length, blood red in colour and possessing a thick body. It is said to resemble the blood filled intestine of a cow, from which its Mongolian name of *allghoi khorkhoi* derives (also given as *olgoj chorchoj* and as *хорхой* in the original Mongolian). Probably one of the best descriptions is that of a leading authority on this cryptid, Czech author and explorer Ivan Mackerle, who describes it thusly[1]:

"Sausage-like worm over half a metre (20 inches) long, and thick as a man's arm, resembling the intestine of cattle. Its tail is short, as [if] it were cut off, but not tapered. It is difficult to tell its head from its tail because it has no visible eyes, nostrils or mouth. Its colour is dark red, like blood or salami... It moves in odd ways – either it rolls around or squirms sideways, sweeping its way about. It lives in desolate sand dunes and in the hot valleys of the Gobi desert with saxaul plants underground. It is possible to see it only during the hottest months of the year, June and July; later it burrows into the sand and sleeps. It gets out on the ground mainly after the rain, when the ground is wet. It is dangerous, because it can kill people and animals instantly at a range of several metres."

The death worm appears to be highly aggressive and is described as possessing two main attack mechanisms; chemical and electrical. Its chemical attack mechanism is said to be based on highly corrosive, yellow coloured venom which it is said to be capable of projecting from the rostral end of its body over some considerable distances, also it is said to secrete this venom through its skin; contact with which is supposed to cause rapid death. The electrical attack mechanism also seems to be long range in nature – being able to incapacitate over distances of several meters.

Sightings, reports and expeditions.

The first English language reference to the death worm was made by the Palaeontologist and explorer Roy Chapman Andrews in his 1926 book *On the Trail of Ancient Man*. He first heard of it whilst at a gathering of Mongolian officials; he describes how[2]:

"None of those present ever had seen the creature, but they all firmly believed in its existence and described it minutely."

Little reference to this cryptid was made for many decades and it languished in obscurity until 1996 when two articles appeared which bought it into the cryptozoological limelight for the first time. The first was an article by Mackerle and featured in *Fate Magazine,* it was in this article that the electrical capabilities of the death worm were first revealed[3]. The second article of that year featured in Karl Shuker's book *The Unexplained* and was elaborated upon a year later with an extensively detailed follow up paper in *Fortean Studies,* which was eventually republished in Shuker's 2003 book *The Beasts That Hide From Men*[4].

It was shortly after this that fresh reports began to surface, such as this remarkable account which was relayed to the *Fortean Times* adventurer Adam Davies[5]:

"He (a local man who assisted some geologists on an expedition) *recalled how one of the geologists had first spotted a khorkhoi and asked him exactly what the creature was. Both of them had been nervous and left the creature alone. The following day, the survey team saw two more in the same area – which they had promptly burned... When I asked the old man why, he turned towards me, stared me straight in the eye, and said: "Because we were afraid. I am afraid of this place. It is a bad place. Since then, I have only been back there twice in my life. You will see..."*

Figure 1: An artistic rendering of the death worm

There are two possible reasons why the death worm remained an obscure cryptid for so long. On the one hand its purported abilities stretch the imagination somewhat and even cryptozoologists, who generally have a reputation for being receptive to 'strangeness', were probably highly sceptical when regarding the case for this cryptid; especially when considered in the light of an evidence-based integrative framework (about which more will be said later). On the other hand there also exists no evidence of a non-sightings nature. There are no carcasses, no photographs – purported or other wise, no in-hand evidence of any kind, which makes the death worm irregular even by cryptid standards. This of course added to the general air of scepticism that surrounded it.

To date, two major expeditions (in addition to Mackerle's long term field work) have been mounted in an attempt to penetrate the Gobi desert and find evidence for the existence of these creatures. The first was a 2005 expedition mounted by the Centre for Fortean Zoology; the second was featured in the reality television series "Destination Truth", produced by the Mandt Brothers which took place between 2006 and 2007. Neither the expeditions nor Mackerle produced evidence supporting the existence of these creatures. It must be noted however that none of these expeditions have been able to survey the prohibited areas along the Mongolian/Chinese border, and that researchers at the Centre for Fortean Zoology have suggested that no conclusive statements can be made about the existence of these creatures until these areas are thoroughly investigated[6].

Figure 2: CFZ Operation Death Worm 2005 expedition patch © The Centre for Fortean Zoology.

Making a case for existence.

i) The plausibility method vs. the integrative method.

Traditionally, cryptozoologists have adopted what could be termed an 'integrative method' for assessing the existence of a particular cryptid. The integrative method typically involves taking a body of supporting data and using it to build identity theories by extracting correlations from the data. The strength of a particular cryptid identity theory is therefore simply a function of a) the amount of reliable data available and b) the number of significant points of correlation within the data. Heuvelmans can be called the father of the integrative method in cryptozoology as he would take reliable sighting and other data and try to extract from them general points of correlation from which identity taxonomies could be derived – as perhaps best illustrated by the technique he used to derive the various sea serpent categories in his 1965 book, *In the Wake of the Sea-Serpents*[7].

An alternative approach developed by this author for assessing the existence of cryptids is what could be termed the 'plausibility method' which is in many respects the inverse of Heuvelmans' integrative method. The plausibility method starts with the development of an identity theory for a particular cryptid which is based primarily upon ecological inference and the suggestion of plausible evolutionary narratives. The identity theory is then compared against what data may exist and re-evaluated in the context of new data via a kind of dynamic feedback that ultimately aims to correlate the extrinsic factors (ecological, behavioural, evolutionary etc. inference) that went into formulating the identity theory with intrinsic factors (sightings data etc.). In order to establish an identity theory using the plausibility method very little data need be known initially, perhaps only enough to establish the geographic location of the cryptid and some basic facts about its purported morphology etc. Occam's razor plays perhaps the biggest part in the initial identity determination as the identity should be plausible above all else until new data can allow for that identity theory to be revised. The integrative method, as has been mentioned, typically requires large volumes of data to be present from which speculative cryptid identities can be extracted by means of correlation, which means that it is useless for handling lesser known cryptids like the Mongolian death worm where there currently exist very little data. The plausibility method however would seem to be much more flexible for assessing these less well known cryptids, as putative identity theories can be constructed and new research direction suggested, using only the available data which can then await falsification or modification in light of any new data.

ii) Previously proposed identity theories.

There have been relatively few attempts made at suggesting an identity for the death worm. This is probably at least in part due to the paucity of data and the corresponding lack of interest in this cryptid. However some have made attempts at suggesting an identity, albeit in a somewhat haphazard fashion in some cases. The identity theories can loosely be divided into two categories; invertebrate and vertebrate.

iii) The invertebrate theory.

The idea that the death worm is some kind of unknown invertebrate type is perhaps the most commonly held view amongst researchers like Mackerle. It features in the majority of graphic representations of this cryptid (see figure 1). There has been virtually no speculation as to what type of invertebrate the death worm may be, although the representations seem to depict it as being eyeless, tubular and segmented which seem to point towards a possible Annelid identity.

iv) The vertebrate theories.

Generally these theories revolve around the idea that the death worm is some kind of snake or another kind of reptile. The argument has been made that the venom spitting capability points towards a possible serpent identity as certain cobras are capable of projecting venom at potential predators[8]. Cryptozoologist Richard Freeman, who participated in the 2005 Centre for Fortean Zoology expedition, has suggested that the death worm may be a large representative of the amphisbaenia or worm lizards – a relatively little studied and ancient family of animals[6]. The idea that it may also be a terrestrial version of the electric eel has also been suggested[8].

Evaluating these identity theories.

Suggesting an identity theory for the death worm is not an easy task. The aforementioned lack of all but the most nebulous of sightings data leaves very little to go on when building a case for any particular theory. The only invertebrates that could possibly be plausible candidates for the identity of the death worm are the Annelida, as has already been mentioned. The polychaete worms are the only members of this phylum known to be equipped with the capacity to produce venom. Some polychaete worms have fangs located at the rostral end of their bodies with which they can impale prey and deter predators, some are also capable of producing venom which can be discharged via contact with the bristles along their bodies. Although some polychaete worms can attain large sizes (9 feet or more in some cases), all known polychaete worms are adapted for a solely aquatic existence. Exposure to dry air would cause rapid desiccation and death[9].

The idea that the death worm could be a fish such as the electric eel, only adapted for a terrestrial mode of existence can be largely ruled out with the observation that there are no known terrestrial counterparts to these fish and that there existence is highly unlikely due to the highly aquatic adaptations exhibited by all electrogenic fish.

The capacity to discharge electricity on land in these cryptids has to be treated most sceptically for two main reasons; firstly the task of delivering electrical discharges is made simple for fish by the fact that water is a conductor; whereas air is a dielectric medium, which would tend to resist the passage of electrical current. This means that vast amounts of energy would have to be generated by the cryptid to transmit an electrical current to its intended victim – orders of magnitude more than any biological system is known to be capable of generating.

Secondly, this purported ability seems to be a relatively recent addition to the mythos surrounding this creature, as it was first revealed to the English speaking world in Mackerle's 1996 *FATE* article, and seems to be all but absent from earlier reports such as that of Andrews. With all due respect to Mackerle, it is possible that this claim could have resulted from exaggeration on the part of witnesses, mistranslation etc. At any rate, there is a strong case to be made for ignoring purported electrogenic capabilities in formulating identity theories for this cryptid.

The reptile theories are more plausible because there are reptiles such as snakes and legless lizards that are known to be tolerant to desert conditions, also, as has been mentioned certain snakes have the ability to spit venom at potential predators, which is analogous to the capacity exhibited by the death worm. However there are two problems with a reptile identity; on the one hand the death worm is described as being venomous to the touch, which would imply an ability to secrete venom through its skin – a non-reptile trait.

Also the behavioural ecology data on this creature indicate a limited activity cycle centred on the availability of water. It is said that the death worm is primarily active in June and July, and tends to surface only after heavy rain. The rest of the year it remains in a state of dormancy underground. This too is irregular for a reptile and seems to be behaviour more suited to an amphibian, perhaps a salamander.

The death worm as a possible salamander species.

Salamanders constitute a very ancient order of amphibians - the caudata. This order is subdivided into three suborders which include the cryptobranchoidea (giant salamanders), the salamandroidea (advanced salamanders) and the sirenoidea (sirens). There exist a great diversity of morphologies and life strategies exhibited by the caudata. The Japanese giant salamander can grow to over 6 feet in length (longer than the purported 2 - 5 feet length range for the death worm); salamanders can also be exceptionally long lived – with the Japanese giant salamander being the longest lived at around 80 years. Some, such as members of the amphiumidae family and the sirens, have worm or eel – like bodies. Many species of salamander are capable of secreting toxins through their skin and in their saliva, and some species, such as the marbled salamander spend considerable portions of their lives underground, only emerging when conditions are right[10].

Much of the death worm's inferred behavioural ecology seems compatible with what is known about the Salamanders.

i) The putative ecology of the death worm.

The Gobi desert is classified as a cold desert and its moisture content varies considerably by season. Approximately 194 millimetres of rain falls per annum in the Gobi, however additional moisture comes from the Siberian Steppes and it is not uncommon to find frost and snow in the Gobi, especially in the winter season where temperatures can drop as low as -40 degrees Celsius. The southern eco-regions of the Gobi sometimes experience monsoon conditions and tend to possess bodies of water[11]. It is possible that the death worm breeds in the moist regions of the south and then migrates into the hotter, dryer eco-regions.

As a salamander the death worm would possess a very cryptic lifestyle, preferring to remain hidden underground throughout most of the year, except during the summer months when it will occasionally surface due to the rains. Its relatively large size would probably allow it to live a very long life, which would be advantageous as it could better adapt to an environment in which access to significant amounts of moisture is rare. It is said also to have a predilection for a parasitic flowering plant called the Goyo, which it may consume[1]. Salamanders in general are known to be highly tolerant of arid ecoregions, provided they have periodic access to moisture; this coupled with the fact that some salamander species are in fact exclusively terrestrial in their adult forms, and in some cases herbivorous, makes the case for a possible salamander identity for the death worm even more compelling.

ii) A speculative evolutionary historical narrative.

If death worms are in fact salamanders, they are amongst the largest, so it is possible that they could perhaps be legless members of the cryptobranchoidea suborder, or perhaps large specialist, terrestrial adapted members of the salamandroidea. However, new studies on the origin of salamanders would seem to suggest something entirely different. The main event that led to the current diversity of the salamanders is thought to have taken place in Asia around 110 million years ago. As the Indian subcontinent began to collide with Asia, the Himalayas and Tibetan plateau rose which in turn created the Gobi desert. Researchers now think that it was this event that split the ancestral salamander populations and forced their adaptive radiation[12].

What if a population of the ancestral salamanders became trapped in the newly forming Gobi desert and gradually adapted to the changing conditions adopting a highly water – conservative life strategy incorporating prolonged dormancy periods and herbivory?

Additionally, what if in response to very low population levels and very long life expectancies, these salamanders developed a highly effective method for dealing with potential predators in the form of a lethal toxin which could be projected over a distance?

The answer to these questions is that something like the death worm could have evolved under these conditions - something that was very rarely seen, and at a glance wouldn't look much like a salamander, but much like other salamanders, would possess a more than ample chemical defence mechanism only with modifications specifically geared to the protection of a potentially very small gene pool.

Conclusions.

It is obviously very difficult to establish with any certainty the identity of a lesser known cryptid such as the Mongolian death worm for which the evidence is very scarce.

However, the advantages of the plausibility method should now be obvious. In looking at the extrinsic ecological factors that have been considered in the context of a possible salamander identity for the death worm, new research targets can be suggested, such as the idea that as salamanders, moister climates would significantly factor into the speculative reproduction of the death worm, which in turn reinforces the view that certain areas of the Gobi desert in the southern regions along the prohibited zones of the Mongolian/Chinese border would be ideal places to search for this cryptid. Obtaining permission to investigate these areas should be the priority goal of all future research into the existence of this cryptid.

References.

1. Mackerle, I. (1992). *Tajemstvi prazskeho Golema* (1st edition). Magnet Press, Czechoslovakia.
2. Andrews, R. C. (1935). *On the Trail of Ancient Man.* Garden City Publishing Company, New York.
3. Mackerle, I. (1996). "In search of the killer worm." *FATE Magazine,* 49:6, 555.
4. Shuker, K. P. N. (2003). *The Beasts That Hide From Man.* Paraview Press, New York.
5. Davies, A, (2004). "Death Worm!" *Fortean Times,* available at: http://www.forteantimes.com/features/articles/158/death_worm.html retrieved on 18/10/07.
6. Freeman, R. "Expedition report: men and monsters in Mongolia." *The Centre for Fortean Zoology,* available at: http://www.cfz.org.uk/dwrep.htm retrieved on 18/10/07.
7. Heuvelmans, B. *In the Wake of the Sea-Serpents.* Hill and Wang, New York.
8. "The Mongolian death worm." Available at: http://www.virtuescience.com/mongolian-death-worm.html retrieved on 18/10/07.
9. Rouse, G. W. and Pleijel, F. *Polychaetes.* Oxford University Press, New York.
10. Indiviglio, F. (1997). *Newts and Salamanders (Complete Pet Owner's Manuals).* Barron's Educational Series, New York.
11. Man, J. (1997). *Gobi: Tracking the Desert.* Yale University Press, New Haven.
12. Zhang, P., Chen, Y-Q., Zhou, H., Liu, Y-F., Wang, X-L., Papenfuss, T. J., Wake, D. B and Qu, L-H. (2006). "Phylogeny, evolution, and biogeography of Asiatic Salamanders (Hynobiidae)." *Proceedings of the National Academy of Sciences.* 103, 7360-7365.

CATS AROUND THE CAPITAL

Neil Arnold

"I stood there, cold, listening to the wind off the night moors. And maybe something else was listening. Something that kills hens and sheep and sometimes men...something that hears you, and comes closer..."
'The Nature Of The Beast' – Janni Howker (1985)

'London Zoo Press Release' – May 8th 2001:

'London Zoo rescues a roaming European lynx from a Golders Green Garden. On Friday 4 May 2001, London Zoo received a call from the Barnet Borough Police based at Colindale Police Station, North London, requesting assistance with a big cat sighting in the Golders Green area.

A member of the public had seen an animal sitting on the wall of her back garden, which she initially thought was a leopard, as it had a spotted coat. London Zoo's Head Keeper of Big Cats, Ray Charter, and a colleague, Terry Marsh, were driven with a police escort to a residential area in Golders Green, where the cat had allegedly been seen in the large garden.

"We get numerous calls at London Zoo reporting big cat sightings and so far all of them have proved incorrect – it usually turns out to be a large domestic cat", commented Ray Charter. *"...so you can imagine my surprise when I bent down to look under the hedge expecting to see a large ginger Tom, only to be met by a much more exotic face!".*

After several attempts to catch the cat with a hand net in the large open area, it was finally contained in a smaller area under some steps of a nearby flat. Having assessed the situation, Ray decided to call London Zoo's Senior Veterinary Officer, Tony Sainsbury, who sedated the animal with a blowpipe. Once sedated, the animal was given a veterinary examination and was found to be a female European lynx of approximately 18 months.

"The lynx was underweight, but in a fair condition", says Tony Sainsbury.
"She is currently recovering in our hospital and we will do a full veterinary examination in the next couple of days. She seems to have a problem with her left hind leg which we will examine under anaesthetic."

The origin of the animal is still unknown.

"It is difficult to speculate where the animal came from", said Nick Lindsay, Senior Curator for London Zoo and Whipsnade Wild Animal Park. *"In order to own an exotic cat species you are required to have a Dangerous Wild Animal License from your local authority."*

There was some concern from local residents regarding the danger that was posed by this animal.
"If left alone it is unlikely that the animal would have harmed a person", continues Nick Lindsay. *"However, if it felt threatened or cornered it could give a nasty scratch or bite. It was more likely to be frightened than dangerous."*

DI Paul Anstee from the Barnet Borough Police says, *"The Police are extremely grateful that they had the back up of London Zoo's expertise in dealing with this unusual event. The animal will remain in the care of London Zoo while she recuperates and her future is decided."*

This Is Local London - September 2001

'Four months after being captured in a suburban back garden, the 'Beast of Barnet' is alive and well at London Zoo.

A wild cat dubbed the 'Beast of Barnet', found in a Childs Hill garden back in May, has recovered well, according to London Zoo.

Times Group readers may recall how Potters Bar and South Mimms residents were forced to lock themselves in their own homes in September 1998 as police combed the area looking for a large catlike animal which had been sighted there earlier. There were also sightings of another 'Beast of Barnet' in the Bookman's Park area.

However, the captive lynx seems to be enjoying her new home, the big cat enclosure at London Zoo having recovered fully from her ordeal.

Head keeper of London Zoo's big cat section, Ray Charter, said: *"We've called her Lara and she's the only lynx in the big cat enclosure here so we will not be breeding her. She's in good condition now."*

RSPCA inspector Dermot Murphy added: *"We haven't been able to find out who owned the lynx but believe it was privately owned. Unfortunately, there's been an increase in people owning exotic animals and we would warn anyone hoping to do this to think carefully as they are not ideal pets in the home."*

The inspector said it is an offence to hold an animal of this kind without license under the Government enforced Dangerous Wild Animals Act.'

DANGEROUS WILD ANIMALS ACT 1976

Conditions subject to which the Licence is Granted.

1. While any animal concerned is being kept only under the authority of the Licence:

a)   The animal shall be kept by no person other than the person specified above.

b)   The animal shall normally be held at the premises specified on the licence.

c)   The animals shall not be moved from those premises without authorisation by the Borough Environmental Health Officer and shall not be moved except for the purpose of veterinary attention or to:

- ☐   A zoological garden
- ☐   A circus
- ☐   Premises licensed as a pet shop under the Pet Animals Act 1951
- ☐   A place registered pursuant to the Cruelty to Animals Act 1876 for the purpose of performing experiments.

Premises licensed to keep such animals under the provisions of the Dangerous Wild Animals Act 1976

d) The person to whom the Licence is granted shall hold a current insurance policy which insures him against liability for any damage which may be caused by the animals and the terms of such policy shall be approved by the Council.

2. The species and number of animals of each species which may be kept under the authority of the Licence shall be restricted to those specified on the licence.

3. Where snakes are kept the applicant shall provide to the satisfaction of the Borough Environmental Health Officer snake-proof wire screens to the window used for the ventilation of and the door to the room in which the animals are kept.

4. The animals shall be kept in containers constructed of materials and be so designed that they will not burn, shatter or collapse when involved in fire to the extent where animals could escape.

5. A properly constructed notice bearing the words 'WARNING – DANGEROUS ANIMALS' in 2 inch high block letters on a conspicuous background shall be securely affixed to the outside of the door of the room in which the animals are kept.

6. The door of the room in which the animals are kept shall be securely locked closed at all times when the animals are not being attended to.

7. In the case of snakes an antidote to the venom of the snakes shall be provided by the person to whom the Licence is granted and shall readily be available at all times including during transportation of the animals.

8. Transportation

a) Transportation of the animal shall only be undertaken in an approved stoutly made container so constructed that it will not shatter or collapse in the event of impact or fire and at all times secures that the animal shall not escape.

b) A properly constructed notice bearing the words 'WARNING – DANGEROUS ANIMALS' in one inch high block letters on a conspicuous background shall be securely affixed to the outside of the container in which the animal is kept.

c) Details of species of animal, availability of the antidote and the names and addresses of the consignor and consignee must be clearly displayed on the container.

d) The Licence holder shall ensure that all the above transportation requirements are complied with if the transportation of the animal is undertaken by a third person other than the consignor or consignee.

Despite the initial excitement of the lynx, many researchers concluded, and rapidly, that this particular animal had only been roaming free for a few days, and clearly wasn't an elusive 'big cat' caught by methodical research or even chance. Lara, had simply been sitting on a garden wall waiting to be fed, and only took off after two attempts to dart her, and one sharp jab to her rear end. Some experts, including Mike Thomas of

Newquay Zoo, believed that the lynx had been injured by a vehicle, which had damaged the poor animals leg. It is very likely that the emaciated cat had only been feeding off scraps for a few days to get by.

In an excellent little article written by Paul Crowther, entitled *'Lara The Lynx Of London Town'*, which appeared in the Newsfile Xtra section of *Animals & Men* magazine, issue 24, he asked, *"Why did it take so long for anyone to report seeing it ?",* and added, *"The idea that a lynx could roam around London for approximately five days without anyone noticing it does beggar belief. If a lynx can roam around London without detection by the numerous people who live and work there, who is to deny that a similar exotic cat cannot survive and go unnoticed on Bodmin Moor for years!"*

The lynx was eventually transferred to Amneville Zoo, in France.

The Eurasian lynx (*Felis lynx*) inhabits Northern Europe and East Asia, although many have been wiped out by hunters and fur trappers. By the mid-20th Century numbers have been greatly reduced, as have the forests which they inhabit. They can reach to over four-feet in length and weigh up to eighty-five pounds. They were once indigenous to the Britain, but it is generally accepted that they died out around four thousand years ago, although bones have been found in caves in Scotland which indicate that some of these animals lived here up until fifteen-hundred years ago.

Scepticism arises to the possibility that such animals still linger on in the present day - however, I do not think this is impossible. Four thousand years is a very long time ago, but for an animal as elusive as the lynx, it's difficult to say exactly when it did die out, if it did at all. Some experts argue as to how these creatures eventually perished in the wilds of Britain. Some claim that the lynx simply died out due to climate change, but others prefer the theory that concerns man. It has been suggested that the lynx began to take the livestock from man mainly due to the fact that man wiped out much of the lynx territory through deforestation. In turn, the lynx would come into farms and kill goats and sheep, which in turn angered man who destroyed the cat.

These beautiful felids can inhabit rocky slopes, forest and open grassland and even though such a cat can kill something up to four times its own size, in the United Kingdom, such a cat would feed on rodents, birds and rabbits. These creatures are easily recognisable by their short tails and tufted ears. Their coat can be of varied patterns, occasionally plain, but mostly rusty or greyish with a mottled effect.

In his book *The Smaller Mystery Carnivores Of The West Country*, cryptozoologist and friend Jonathan Downes wrote of several lynx sightings dating back a few centuries.

He wrote: *"...reports of lynx-like animals in Britain are not a modern phenomenon. The 16th Century author Ralph Holinshead wrote, 'Lions we had very many in the north parts of Scotland and those with manes of no less force than those of Mauretania; but how and when they were destroyed I do not yet read'. However, the most quoted refer-*

ence to an early lynx sighting in Britain, came from the pen of William Cobbet, who was once described by English writer Sir John Verney as, "...farmer, writer, political commentator...", in his Rural Rides published in 1830. It is said that as a boy he observed a grey cat as big as a middle-sized spaniel dog whilst visiting Waverley Abbey in Surrey with his son on the 27th October 1825. One hundred and thirteen years later, on 19th March 1938, also in Surrey, this time at Lightwater, an Irene Roberts wrote a letter to The Field magazine, to speak of the strange cries she was hearing outside her bedroom window of a night. Some of the cries, which she heard during the early hours of one July day, in 1937, were described as "...of peculiar intensity, expressing, it seemed, mortal fear and physical pain".

However, Irene seemed quite knowledgeable of the sounds made by foxes or a rabbit being killed, but attributed these cries as from an unknown animal. Little did Irene realise that a quarter of a century later, these cries would manifest time and time again across the fields of Surrey, and many witnesses would come forward to report what was to become known as the 'Surrey Puma', the first 'big cat' headline.

The puma (*Felis concolor*), also known as the mountain lion, cougar, and in parts of the America, the panther, is the largest of the lesser cats, which also include caracal, Canadian lynx, Eurasian lynx, jungle cat, bobcat and leopard cat. It is not a 'big cat', like the leopard, lion, tiger, and cheetah for it cannot roar, but instead lets out an eerie scream, which can travel for many miles across a valley and the rocky climbs it roams in North, South and Central America. In some parts of the U.S.A. this animal is extremely rare, and has taken on the form of a ghostly, mythical felid, whilst in other areas, such as New Jersey, many sceptics argue against its existence, believing it was wiped from the map a century previous, despite modern days sightings of what the locals call the 'shadow' or 'ghost' cat. Rather strangely, there were probably more reports in Surrey, England, during the 1960s, than in a whole century in its native country!

In the United Kingdom, the press often confuse the puma with the black 'panther', although the black 'panther' is simply a melanistic leopard, a darker form of the 'normal' leopard, which is of no relation to the dark-tan coloured puma. The black puma is phenomenally rare, in fact only one specimen has ever been caught and that particular individual was debated. In parts of the United States many eye-witnesses are convinced that the black animals they are seeing are not black leopards, but melanistic, screaming pumas, but until one is captured there will always be a strong degree of scepticism, especially when you consider that in many areas of the U.S.A. the puma is said to no longer exist, despite persistent reports, but for sceptics and scientists to accept normal mountain lions is hard enough, so for them to believe in black specimen's is nigh on impossible. Some specimens in the United States have been found with very dark tan coats, but not black, or extremely dark as to look black, like that of the melanistic leopard. However, when a majority of the early reports of 'big cats' emerged in Britain during the middle to late 1900s, much confusion arose regarding species identification. Much of the public, which is very understandable, were unaware that leopards could be

-CFZ YEARBOOK 2008-

black, and so if you take a stroll through many press reports, even today, regarding black cat sightings, many reporters will still describe witnesses as seeing a 'black, puma-like cat', instead of the black leopard, which we will talk more of later.

The puma is a reasonably large cat, and can reach over six-feet in length. The young of the puma have spotted coats, - these spots gradually fade out. The tail of the puma, which is a wonderful balancing tool, can measure up to thirty inches. In parts of the U.S.A. they are feared as a threat to humans, although bees and dogs cause more deaths each year.

On the 11th January 2004 *The Sunday Mirror* newspaper reported, '*Lion Kills A Cyclist*', and the story was echoed in *The People*, under the headline '*Girl Saved From Puma*', and also spread like wildfire across parts of the U.S.A. where the incident took place. *The Mirror* reported that, "*...a mountain lion was shot after mauling a man to death and dragging a woman one-hundred yards with her head in its mouth. Rangers hid near the body of cyclist Mark Reynolds, 35, waited for the lion to return to its prey and then killed it.*"

The beast also attacked Anne Hjelle, 30, a fitness instructor. It pounced on her as she cycled with her friend, Deborah Nichols, in parkland outside Los Angeles. It sank its jaws into her head and dragged her one-hundred yards into a bush while Miss Nichols held on to her friend's legs. Miss Hjelle was in a serious condition in a LA hospital. Describing the attack, Miss Nichols said: '*I held onto her and kept screaming and screaming. The lion just wouldn't let go of her face.*'

One of the first ever recorded reports of the Surrey puma came from 1955 when a female witness, walking her dog near Abinger Hammer, in Surrey, discovered the grisly, half-eaten remains of a calf. She claimed, that after finding the gory mess, she was shocked to see a puma-like animal slinking out of sight. Four years after the sighting, during '59, a Mr Burningham, driving near Preston Candover in Hampshire, saw a large cat cross the road in front of his vehicle. He described the animal as being yellowish in colour, and the size of a Labrador dog, but with a cats head. Its coat looked quite rough and its tail was long. Thankfully, the witness wasn't too dumbstruck to slow his vehicle and watch the animal for a short while as it observed sheep in the adjacent field to the hedgerow it was sitting in.

That same year a taxi-driver described seeing a 'lion', which leapt over a hedge in the vicinity of Tweezledown Racecourse, but things really began hotting up during the early '60s, when a huge flood of sightings bombarded the local newspapers, literally forcing the news crews and media hounds to wake up and take notice that some 'thing' was indeed out there stalking the fields. However, although 1963 is seen as the defining year in which the Surrey felid made its name, there were still handfuls of previous reports lurking around as anecdotal evidence.

A now defunct Surrey related website page reported, on the '*Scare Bears & Other Creatures*' around Surrey, and mentioned an intriguing report from 1961:

'*In 1961 golfers spotted a 'big black animal' (the first ever Surrey black leopard sighting I wonder ?) in the autumn mist on Croham Hurst Gold Course. One man bravely moved closer to get a better view. He judged it to be a bear about three-feet tall. It disappeared into the woods as the friends peered at it across the field. The golf club feared the animal might scare women and children, so police were called out, but a search revealed no trace of the mystery animal. Officers suggested it could have been a large dog, while Saint Bernard's or dog badgers were put forward as solutions by some locals.*'

Even more interestingly, the website went on to mention a 'big cat' escaping in Surrey, very early evidence that cats were present in the county, whether roaming free or circus bound.

'*In the 1920s 'Carmo Manor' in Shirley acted as the winter quarters of Carmo's Circus. The Great Carmo's menagerie was housed there, and the circus men would often wash the elephants in the old estate pond or take bears for exercise round the grounds. So when a woman rang the police station to report an escaped leopard the duty sergeant was on the point of organising a full-scale, armed leopard hunt. The creature had been seen to force its way through a hedge and then jump over a fence. Stopping a moment to check the facts, he rang Carmo's and found they had no leopards! This time, the twilight at dusk was accused of turning a Dalmatian into a vision of a leopard. The circus dog had apparently slipped out during training for a new act.*'

It would be nice to prove these details as correct and that it was indeed the circus dog that had escaped. One of the more popular theories on these wild cats roaming the British Isles is that they are indeed generations spawned from cats that once escaped from the local circus or private collections. The Romans also had menageries dating back several centuries, which we'll speak of later, and these would include large cats such as the leopard. Strangely though, the normal spotted leopard, known for its rosette-marked coat, is rarely sighted in Britain, but melanistic leopards are extremely common.

Why is this so ?

Well, firstly, there is no difference between the two, they are the same animal, native to Africa and Asia. The black 'panther' is simply the name given to the darker form of the leopard. This skin pigment, known as melanin, is a common condition. Two normal parents can produce up to four young, and within this litter there can be melanistic individuals. However, melanistic parents can only produce melanistic young, which is why reports of leopards in Britain only concern dark-coated cats. These cats are in fact very dark brown, but from a distance look black. If these individuals are seen at closer quarters and under the right light conditions, the rosette pattern can be seen. However, there are an increasing number of eye-witness who describe what they believe are jet-black cats. A recessive gene in the melanistic leopard could in fact gradually phase out the rosette pattern, or increase the darkness of the coat to an almost jet-black colouration. It's clear that in a majority of sightings regarding black cats, people are not merely seeing large feral cats, dogs or any other animal. They describe an animal able to stash prey up in trees. An animal that can reach over four-feet in length, be of muscular build, and omit a deep, sawing cough noise and double-barrel snarl.

The leopard, in its country of origin, is feared as a man-eater but much of this is myth. In Britain there is no evidence to suggest these animals are a danger to the public unless injured, provoked or cornered and so far, the attacks that have allegedly taken place, involving humans, are dubious at that. Some researchers argue that there are a selection of mystery black cat sightings which could point to a new species of British cat, something smaller than the leopard, but reports seem inconsistent, and also in regards to the possibility that such sightings of mystery British cats are unknown species. Zoologist Karl Shuker wrote:

"...I would be only too delighted and thrilled if a totally new species of large felid were to be discovered in Britain, as a zoologist surveying the possibility realistically my personal opinion of this is that its chances of success are gravely hampered by basic problems...To begin with, the British Big Cat would need to be nothing short of a shape shifter to account for the immense variety of felids reported – cats which are black, grey, every shade of brown, striped, spotted, with long legs, with short legs, with long tails, with short tails, with ear-tufts, without ear-tufts, which roar, which scream, which are of small dog size, which equal the size of the largest dog breeds etc, etc. Certainly no single non-domestic species of felid (indeed no wild mammal of any single species, known or unknown) could possibly exhibit so many markedly different forms...".

Maybe this next fact is press influenced, but it seems that a majority of Surrey puma sightings were of fawn-coloured animals. Was this because quite simply there were hardly any melanistic leopards around Surrey at the time, or because a hysterical public were only spotting the puma simply because the media had only informed them of such a cat, without ever mentioning other species?

Mr Ernest Jellett, who worked for the Mid-Wessex Water Board, was cycling near Heathy Park Reservoir at around 7:45 am on the day of 16th July 1962 when he saw a big cat with a flat face and big paws stalking a rabbit. Although at the time Mr Jellett described the creature he saw as something akin to a 'young lion cub', it seems clear that what he did in fact see was a small puma. In the December of '62 another water board employee saw the cat, and there were numerous other reports of a rabbit-coloured cat prowling the woods. Now, at the time, scepticism was rife. A majority of reports, even today are not reported by some witnesses due to fear of ridicule, but there is more of an understanding and coverage of these sightings, with various research groups being set up to record the evidence of these exotic cats.

During the 1960s, in fact, right up until the 1980s, a majority of sightings were not taken seriously. What didn't help was the fact that witnesses did not know what they were seeing, but this was of course natural. Nobody cycling down a country lane in Surrey expects to see a large cat, a big, slinking felid which they cannot recognise simply because all they are used to seeing are foxes and badgers. So, this is where many of the 'lion' and 'young lioness' reports are born from, but a majority of these must account for the puma. Why do I know this? Well, quite simply put, there are no tigers, cheetahs or lions roaming the U.K., despite some of the more modern panics that have circulated, such as the Nottingham 'lion' scare, which took place in the 1970s, and others. In fact, the only time such

an animal would take to the countryside was if such a big cat had escaped from a zoo or a circus. Even then, such animals would be easier to trace, they will seek larger prey, especially the tiger and the lion and also in the case of the lion, prowl open ground whereas the animals we are dealing with in the British Isles are very much elusive and very British exotic felids.

Reports snowballed into late winter, through to 1963, when a mysterious predator paid various visits to Bushylease Farm, between Crondall and Ewshot. Something silent had been spooking the farm dogs, and fleeting sightings of the slinking intruder often described it as tan-coloured. A huge hunt ensued for the creature during the Summer of '63 after a David Back spotted an animal laying by the roadside at 1:00 am on the 18th July, at Shooter's Hill. One-hundred and twenty-six policemen, accompanied by more than twenty dogs, alongside ambulance staff, and officials from the RSPCA, scoured more than eight-hundred acres of land in the hunt for a cat they believed was a cheetah, despite no actual reports from the time describing such a thing! A complete waste of time and man-power, as usual, combed the vicinity of various sightings, but to no avail. And such a ridiculous attempt to flush out a cat was, unfortunately to be repeated time and time again, up to the present day, with the same results, which often leads many to believe that surely, these animals cannot exist if they are so clever to evade hundreds of trackers. Albeit cumbersome and noisy ones at that!

Many of the sightings which emerged from 1964 were chronicled in Di Francis' superb book, *Cat Country*, and some read as follows:

27th September 1964 – 10:45 pm – a puma-like cat was sighted by a motorist at Loxwood. The next day on 28th September 1964 – 6:15 pm – a female witness observed a puma at Witley. Then, on 29th September 1964 – 6:45 am – a road worker in the Puttenham vicinity came face to face with a large cat. There were also a couple of reports of a puma on October 3rd, and then, throughout October consistent eye-witness descriptions from Hascombe, Hindhead and Elstead, all reports describing a puma-like animal.

The following year Di Francis catalogued another consistent batch of reports, with one at Chiddingford where a female witness, tending to her horses, described how a 'big cat' had leapt over her head. During 1966 there were further reports from Puttenham and Godalming also saw several sightings. Unfortunately, Di Francis only chronicles these reports as 'puma-like', so I am assuming that all these reports, which certainly reached over fifty during '66, concerned fawn-tan coloured animals and not black cats. As I have mentioned previous, the press often wrote of these sightings rather loosely, and at the time many journalists would have been unaware of black leopard sightings, and many would have thought that a puma-like animal could well have been black, and so it seems, a majority of these reports, from the early '60s, and on to the 1980s, would have nearly always pertained to the 'puma-like' animals, despite the possibility that other animals were involved, including the lesser known cats such as the golden cat and jungle cat.

Flick through any number of newspaper articles pertaining to cat sightings and you'll notice a degree of ignorance from the journalists regarding identity of species, and so many headlines will only contain references to a 'big cat', which is vague to say the least, and not informative at all.

During 1966 several motorists filed reports of large cat-like animals, especially in the Puttenham area. Many of these reports came from the summertime and into autumn. At Ash Green on the morning of July 14th, a witness saw a big cat run into fields, and the same day, only three hours apart at Worplesdon, a witness observed a puma from just thirty yards away as it padded its way into the garden of the householder.

A puma was reported again during mid-August, this time at Cutt Mill. This time the police were spoken to regarding the sighting, although I'm unaware as to what action was taken. At Wormley, a few days after, another householder observed a large cat in the garden, but at Milford on 4th September 1966, a male witness was dumbstruck when a large cat sprang from a tree and sped off. Not only was this witness unaware that large, exotic cats were roaming the Surrey countryside, but also they lay up in trees, a common trait of the leopard which is a superb climber, where, in their countries of origin, will stash prey in the branches of trees to avoid scavengers. Large cats such as the puma and leopard are agile predators, but despite having no fixed dens, unless to give birth, they will lay up wherever possible. Gardens, trees, churchyards, ruins, under cars…there's nowhere that a 'big cat' will not consider.

Many witnesses to large cats in Britain often describe grisly evidence in the form of half-eaten remains, often stashed up in trees, although smaller prey such as rabbits and pigeons are completely consumed with no real trace except a few feathers and bones. The back legs of lambs have often been discovered in fields, as well as the remains of sheep, domestic cats, small dogs, calves and foxes. Sheep are usually half-eaten, cleanly devoured and rasped by the tongue of a large cat, domestic cats have been found headless, foxes disembowelled, and even horses have been found, although alive, in some distress, bearing scratch marks on their flanks and bite marks to the throat. A large cat will not kill out of spite, or tear like that of a dog, it will kill to survive. Traces of these animals are few and far between, a few carcasses here and there, a handful of dubious pugmarks which, unless very defined in mud or sand, are often smudged on bridle pathways, fields, and tracks, and there are also the scratch marks left on trees which, when found, are extremely impressive, especially when found higher than any badger can reach.

During the September of 1966 many witnesses claimed to have found the paw-prints of big cats. These echoed similar evidence found in 1964 when, on September 7th, P.C. Bill Cooper was called out to investigate a strange set of impressions trailing across a field at Stileman's Racing Stables. The prints seemed to measure five inches across, and suggested that a heavy animal had made them. Unfortunately, after following the prints for half a mile, they disappeared into undergrowth. However, photographs were taken of the prints and examined by experts at London Zoo who confirmed they had indeed been made by a large cat such as a puma. Unfortunately, a sceptical Dr. Maurice Burton suggested that the prints were made by a dog. What we must remember when looking at prints, as mentioned elsewhere, is that the dog does not retract its claws but a large cat will. Claws may be thrown out to grip, but a majority of cat prints can be confirmed by the lack of claw marks. A puma was seen at Stringers Common on the evening of the 22nd September '66 by a woman out walking. There had been two other sightings earlier in the morning at Hog's Back. One of these concerned a motorist who had an impressive sighting of a brown cat which was caught in the headlights as it slinked across the road.

The same animal was seen at Hog's Back in October of '66, again, a motorist had caught the animal as it crossed the road, and before the year was out there'd been another handful of reports and the rest of the decade was no different, with sightings coming from Wood Street and Thursley.

There were claims during the very late 1960s that a puma was shot and killed in the Surrey area, as well as Sussex and Hampshire. However, there is no proof of this, but if such an incident did occur then it is very likely that at the time the carcass would have been buried or burned.

Reports continued into 1968. Farnborough was an extremely active area. However, sometime during 1970 a rather vague critter was said to have been prowling Hackney Marshes, in East London although little appears to have been recorded regarding a creature that was described as being bear-like! Rather quirkily, English comedian Jasper Carrott and actor Robert Powell were to bring this obscure legend to the public eye and out of the depths of obscurity for an episode of their series, '*The Detectives*' in 1997! A paranormal-related website claimed that the animal, which apparently left prints in the snow on one occasion, was active into the early 1980s as well.

Graham McKewan's important book pertaining to strange creatures, entitled, *Mystery Animals Of Britain And Ireland* catalogues many eye-witness sightings touching the borders. One woman, a Mrs Anne Stanette, described her encounter a decade later, in a letter, which happened in East Surrey:

"*While I was out riding in Granger's Woods, Woldingham, in May 1978, I saw what I believe was a lion rush across the road in front of me. It ran from the Oxted side into thick bushes on the opposite side. It was about ten or twelve yards from me. It was a beige/light brown colour and had a small head in comparison with the rest of its body. (It had no shaggy mane).*"

Such a sighting, most definitely of a puma, proves how natural it was an occurrence, back then more so, for witnesses to wrongly identify the puma, as a lioness. At the time, much of the public, despite the spate of Surrey sightings, and also handfuls of sightings across Britain taking place, were familiar with the term 'big cat', but weren't quite sure exactly what 'cat' was roaming the countryside. A majority of animal documentaries, even today, cover the world of the lion. The puma is extremely elusive, and so footage is not as easy to come by, but back in the '60s people seeing beige cats often felt they were seeing a lioness. However, the puma does have a small head in relation to its body, whereas the lioness is a much bigger animal, with a bigger, more squarer head and its most distinguishing feature is its colouration, more of a sandy yellow colour. However, back then, and still, rather surprisingly in the present day, people do not, and did not expect to see a large cat skulking around and when they did spot one, a majority of people were probably unaware of what a puma looked like, unaware also that leopards can be black, unaware of the differing species actually out there, and so in their mind they probably put two and two together and guessed that they'd seen the only animal they really knew, the lion. However, two years previous in Nottingham, several witness came forward to report what they'd described as, "...definitely a lion", roaming heavily populated areas. Even the Nottingham police were alarmed by the sightings and said they were "...ninety-eight percent certain", that some kind of animal was out there. However, after many reports and countless searches, no lion was unearthed, but puma sightings in Nottingham, and other cities across Britain persisted.

From the New Forest area of Hampshire came several reports of a big, black animal, during the early 1970s. John McPherson saw a massive cat cross the road in front of him on January 23rd 1973. In the September of 1976 three lions escaped from a circus at Epsom in Surrey but were recaptured, but not before a horse was severely savaged by one of the large cats. A black cat with a very long tail was also seen in areas of Sussex during the summer of 1979, whilst five years later there were puma reports from Hertfordshire. Thankfully, one or two reports of large, black cats were filtering through to the press, proving that a variety of large animals were roaming the home counties.

Unfortunately, the 1980s proved just as ignorant with regards to the coverage of the sightings. Reports were still consistent, possibly even more so, as any animal that may have been released during the '70s after the introduction of the 1976 Animals Act, would surely be breeding if there were so many cats out there. Sceptics argued that the evidence was scarce which in turn, to them anyway, meant that the animals were scarce too, but this didn't explain the numbers of animals being reported up and down the country. Unfortunately, the sceptics, and some of the journalists truly believed that there were just one or two escaped cats out there to be blamed for the hundreds of sightings which were now coming from parts of Scotland, the West Country, Midlands, Wales, parts of Ireland and down towards Kent and the home counties. The suggestion that just a handful of cats were the cause of all these sightings was quite simply more absurd than the mystery they were pursuing. And yet it still happens today. This is a typical problem with the media. Once an animal becomes the 'beast of Exmoor', or the Surrey puma, the finger is pointed at one animal, when in reality there could be anywhere between five and twenty cats out there, three to four species, yet they all become the local 'beast', something akin to a werewolf movie.

Strangely also, during the 1970s, a naturalist named Maurice Burton concluded that all the reports of alleged large cats, which he'd investigated, turned out to be fox, badger, deer, dog, otter and feral cats. Maybe it was this kind of poor research that turned the 1980s into such a non-event for the Surrey 'big cat', although this lack of activity could also be blamed on another mystery felid, the 'beast' of Exmoor, which literally exploded into the '80s, and became the most popular, headline making 'big cat' in the business. However, reports still existed.

In 1984 at Peaslake, hair samples, believed to have belonged to a puma were analysed, and commented on as proof that a large cat was around Surrey. In the same year there were also sightings at Chiddingford and Witley Park, and in '85 regular reports from Esher, whilst Hertfordshire, Sussex and Kent were also providing several flaps of their own, proof that the animals roaming Surrey were indeed not responsible for the eye-witness reports in the neighbouring counties. During 1987 there were also rumours that a large cat, possibly a puma, was shot and killed near Greenwich Observatory, London.

The 1990s was indeed a popular time for sightings of exotic felids, and the puma of Surrey was often regurgitated when activity arose elsewhere in the U.K. Press interest was slightly more serious by the early to mid-'90s, and certainly more consistent, with reports featuring on the worldwide web, television and in the press on a weekly basis nationwide, and also worldwide, as strange reports of 'out of place' large cats filtered from Australia, the United States of America, Italy, New Zealand, France, Denmark, and even the Isle of Wight!

During the Autumn of 1998, a sandy-coloured cat with a ring on the tip of its tail was sighted at South Mimms, Hertfordshire, and there were several sightings near Potters Bar, after which commenced another fruitless police search. This animal became the 'beast' of Barnet. By this time also, my own research into sightings in Kent and the south-east, had taken on its own snowball effect, with the 'beast' of Bluebell Hill, near Maidstone, being the local favourite, although since the '80s I'd filed reports, and looked into older cases, but by the time the '90s were over, I'd been consistently receiving weekly reports which transformed into a huge cabinet of data, which, when exhaustively studied proved that there was indeed more cats out there than I first thought. Even with siphoning out some of the more dubious and vague reports, annually I was receiving over two-hundred eye-witness accounts of large cats roaming the countryside. Just how many weren't being reported? During 1993, Mr Irvine saw a large cat at Hayes Common in Bromley, one of several reports from the outskirts of London that never reached the press.

Mr Irvine was driving to work one morning at around 6:45 when a charcoal-grey cat, with rosettes bleeding through the coat, bounded across the road ahead just fourteen-feet away. The animal had a smallish head, a very long tail and small ears. At the time Mr Irvine thought this animal may have been an escaped snow leopard. During the 1975 a clouded leopard escaped from Howlett's Zoo in Canterbury, Kent. The cat was at large for eight months before it was shot and killed by a farmer whose lambs had been taken by the animal. However, in 1974 Fred Arnold saw a similar cat cross a road in Folkestone, not too many miles from Canterbury. He too described the animal as greyish in colour, with a yellow tint, and two years previous a Peter Cookson, of Lympne, was driving during the early hours of a June morning when he too saw a yellowish animal, however he described it as having stripes!

Meanwhile, across the Thames – *'Four More Sightings Of The Beast Of Ongar'*, reported the *This Is Local London* website on Saturday 8th August 1998, claiming, *"... witnesses have told police the animal, seen three times in fields at Matchling Green, looked like a young panther. The most recent sighting was off Stanford Rivers Road, Ongar, last Wednesday."*

Essex was certainly already on the map regarding sightings of large cats, but with years of reports coming from Exmoor and Bodmin from the West Country, and also Surrey, many other nationwide reports became either obscured, or simply forgotten.
The previous year in Essex there'd been sightings of a 'big cat' around Wood End, but like so many others, the press merely obliterated any truth by stating the strangest 'facts'. One classic report as follows:

"Commenting on the Matchling Green sightings, PC Chris Caten said: 'they said its tail was panther-ish but it was not as big as they expected a panther to be'."

If this wasn't confusing enough, a local website reported:

"...confirmation that a 'big cat' is stalking the area came in February after an expert examined a goose, one of four savaged in attacks in Weald Bridge Road, North Weald. Claw marks on its body proved it was killed by a lynx or puma-sized animal."

So, we can see by these few sentences just how confusing these cat stories are painted to the public. We have a local policeman claiming that witnesses described to him a 'panther-ish' tail, whatever that is, and that the animal was, "...not as big as they expected a panther to be", even though the report clearly states that the witnesses saw an animal resembling a young panther...which I'm assuming was in fact a black leopard. We also have the confusion of the claw-marks on the body of the savaged geese which an expert claimed were made by a lynx, or a puma-sized animal! Well, which is it to be?

During September 1998, the cat of West Essex became the 'beast' of Bassett, and according to some websites, was identified as a Eurasian lynx, and even investigated on the BBC series *The X-Creatures*, presented by zoologist Chris Packham. He commented:

"About five-thousand years ago they (the lynx) would have been very common in Essex...", although the animal stalking the woods during 1998 was, in his words, *"...an escapee from a zoo or private collector."* Yet here was a man I'd worked with in the past who claimed that animals such as the black leopard did, quite simply, not inhabit our woodlands!

A PC Ross Luke claimed that sightings around Ongar had dated back to the mid-'80s, although there seemed to be some confusion between the sightings, as many reports seemed to describe a large black animal. Whatever the case, Mr Luke claimed that, *"We have a contingency plan; if the cat is ever cornered we would call up the Tactical Firearms Unit and it would be shot...dead."*

During January 2001 a black leopard was sighted at Walthamstow, near marshland. The sleek felid was observed going through bins at 7:45pm one evening by a woman sitting in her car waiting to pick up her son. Despite the fact that the animal came close to the vehicle, the witness described it only as fox-sized, but having a long tail, although in a statement to the press she claimed it was a 'huge animal'. The animal eventually slinked away toward Wickham Close.

On the 31st of January a report was filed concerning a Christmas sighting of the Ongar cat. A Paul Ayton described how he was driving along Greenstead Road when he saw a large black animal which at first he took to be a dog. The creature was walking alongside a hedgerow a few hundred metres away but then headed towards the road. As the animal came closer, via a field, Paul noticed its feline gait. Eventually the witness slowed the car until just twenty yards away from the cat before it slipped away out of sight alongside a house.

Thankfully, as of writing into 2007, these animals have eluded their pursuers and sightings of large cats still continue in Essex. Colchester, Witham, Clacton and Tiptree, have all been areas where large prowling felids have been sighted throughout 2006.

However, the strangest cat flap confusion regarding Essex, and in relation to Kent, took place during the 4th January 1975. According to reports in the *Daily Mail*, and *Weekly News*, an eight-week old 'panther' cub had been stolen from Colchester Zoo. However, the zoo never made the claim until a week later. On the 5th January, angler Fred Lloyd was enjoying a days fishing on the banks of the River Medway at East Peckham when he was disturbed by a noise in the undergrowth. Much to his surprise a two-foot long 'panther' emerged from the bushes. The animal, which Fred took to be a cub, began to hiss at the shocked angler, but instead of facing the fisherman off with prowess and arrogance, the cat could only manage a comical tumble down the bank towards him. In a split second Fred bravely grasped the cat by the scruff of its neck and bundled it into his fishing box. He then packed his gear up and drove home, taking the peculiar find with him. According to reports, the cat ended up in a baby's playpen, which it attempted to destroy with its sharp claws.

After a short while Mr Lloyd phoned around the zoo parks but claimed that the response was of laughter and mockery. It wasn't until a day later that an RSPCA inspector could be bothered to turn up at Fred's house to take a look at the mystery cat that no-one had believed in. Much to his amazement, there it was, a baby Leopard bundle, ready to be taking to a new home. Bizarrely though, Colchester Zoo, then decided that the cat belonged to them and was in fact 'Zar', who was worth something in the region of six-hundred pound. I'm not sure if the mystery cub ended up at Colchester Zoo. At the time, the RSPCA inspector said that before the zoo rang, the leopard was due to end up at Godstone in Surrey. Either way, one question must be asked. How could such an exhausted cat end up more than fifty miles away at East Peckham? Had someone stolen it and then let it go or, did Mr Lloyd have a genuine encounter with a wild, British big cat cub?

'Cat Brutally Murdered' – 5th September 1998. *This Is Local London.*

'A dead cat, with its head cut off, is the latest gruesome find in a series of bizarre attacks on animals across London. The cat was found with its head missing and most of its blood drained, in Main Road, Sidcup, last Thursday. It was removed by Bexley Council environmental health officers.

Several months ago another cat was found in a front garden of a house in Penshurst Avenue, Sidcup. Its head and tail had been removed.

There was also a similar incident in Erith.

A total of forty suspicious deaths among pets have taken place in the last ten months and the most common victims are domestic cats.

RSPCA inspector Nigel Shelton said: "The number of cases of animals which have been decapitated or had limbs removed from their bodies is growing at an alarming rate and we would urge anybody with any information to contact us urgently. An RSPCA spokesman said they had no idea of the motive behind the killings and said the widespread locations of the deaths made it likely that more than one person was involved.

She said police were now following up the leads given by the public through the RSPCA's emergency hotline.

She added: "At the moment, we have no hard facts about the person or people behind these attacks."

With the grisly, but seemingly unrelated cat-rippings making the local headlines, it was the turn of the Barnet 'beast' to rear its ugly head again, this time on September 25th 1998, when two policemen, Martin Stainton and Matthew Durkin, observed a cat which the press claimed was, '…either a puma, cougar or mountain lion', despite the fact that all three of these names belong to the same species of animal! And, in typical police fashion, a helicopter and twelve more policemen arrived on the scene to frighten off the animal.

There were between five and ten sightings of the Barnet prowler during the autumn months, and local inspectors claimed that the animal must have come from a private collection or been released on purpose, despite the nationwide sightings of similar animals.

By the 10th October 1998 the elusive yet harmless cat of Barnet had become known as the 'M25 Monster', and a report in the London press dated 31st October…yes, Halloween, described a *'Woman's Trauma Over Gruesome Cat Murder'*:

'A traumatised sixty-eight year old woman has warned pet owners to be extra vigilant after neighbours found her cat dumped in a garden with its head severed. Devastated pensioner Pamela Stockham of Caxton Road, Wimbledon, had been struggling to come to terms with the gruesome killing which happened sometime between October 15th and the 21st.

Mrs Stockham first became worried when George – her pet of seven years – did not return to her house for twenty-four hours. The following day she targeted nearby roads with a leaflet campaign and put up posters in a bid to find her beloved cat. Just days later a neighbour in Garfield Road broke the news that George's body had been dumped in her garden. His head has not yet been found.

Mrs Stockham told the Guardian: "I am still in absolute shock – I just can't believe it has happened. Who would do such a terrible thing?"

"We know it's George's body because the description matches. He was black with white feet, tummy and neck and had white whiskers. I want to warn everybody who owns cats to be on their guard. The people who do these terrible things should be prosecuted – but right now I just want to kill them."

At the end of November *This Is Local London* claimed in an article that the strange cat killings had spread south, with ten rabbits also suffering the same fate as the many cats. Reports of dead cats circulated around Tottenham, Stepney and New Barnet when a dead cat was discovered on November 11th. Things then became even more serious when at the beginning of the December a psychiatrist was called in to draw up a profile of the London cat-ripper after claims that some forty animals had been slaughtered. A £1,000 reward was put up for the capture of the sadist but by December 19th 1998 someone - or some 'thing' - had struck again, this time in Twickenham, when an Eileen Tattershall lost her cat 'Bonkers'. A neighbour found the dead cat, bereft of head and tail but what was more sickening was how the RSPCA, during the April of 1999, attempted to solve the gory mystery. They claimed, after months of methodical and exhaustive research into the killings, that all of these deaths could be attributed to one killer – the car! An inspector told local press that post-mortems on the decapitated victims had proven that these poor animals had been killed on the road. Whilst other deaths could be attributed to scavengers such as foxes, dogs and even crows!

How the RSPCA came to such a conclusion is beyond me. Unless there are serial killing foxes, killer crows and knife-wielding dogs out there, I really couldn't for the life of me see how the victims could have their heads and tails removed, and the carcasses to be bereft of blood. And how a car manages this is also beyond me, and of course, what kind of car decapitates a domestic cat and then returns it to its home garden? The Met's wildlife liaison officer, Andy Fisher also managed to come up with this classic statement: *"There has always been cats and there has always been traffic. What we don't know is why there has been a sudden increase in reports."* And what about the rabbits?

In the summer of '99 reporter Rob Bailey of the *Dartford Messenger* filed a sighting of a large cat in Orpington, at a area known as Badger's Mount. There had also been reports of large brown, as well as black cats around Swanley, Darenth, up into Dartford and across in Sidcup where some of the domestic cat killings had taken place. All this was starting to take on a sinister yet familiar form? Was there really a twisted human being out there severing the heads of domestic cats, or were these decapitations the work of a feline form? Or both? Also, during this time, populations of cats across Kent, the outskirts of London, Sussex, Essex and Surrey were on the rise. A handful of large cats would soon become known as the Bluewater 'big cats', or as one newspaper dubbed them, the 'beasts' of Bean, whilst these territories would also cover Gravesend, in Kent, Sevenoaks, and further down in Tonbridge. It was difficult all of a sudden to monitor individual animals, there was an explosion to look at, a snowball effect that had been clouded by the media and their ability to confine certain cats to certain areas, when the fact of the matter was, there was an abundant population of varying species across the south.

On the 6th February 2001, a report was filed by the *Dartford Messenger* concerning a cat sighting near Bean, on the B255. A large animal leapt from a bank as Christine and Raymond Pearson approached in their vehicle. They described it as fawn in colour and around three-feet long. However, sceptical Derek Gow of the Herne Bay Wild Wood Discovery Park stated:

"If there was a puma living in the local countryside there would be distinct signs, such as traces of their scent."

What I found unusual about this statement was the fact that it's such a disregard of the intelligence of the public. People who see these animals are often unaware of what species of cat they are seeing, let alone the scent or signs they leave, for someone to be so dismissive on the grounds of lack of such obscure evidence is ridiculous.

On the 17th August 2001 the cat beheadings hit the headlines again. For more than a year things had gone a little quiet despite the fact that large cats were still being sighted at a record rate across the country, and many of these sightings had taken place in areas where domestic cats had previously been killed.

The head of a nineteen-year old white cat was found in a back garden in Penge. Shortly after this incident the half-eaten remains of a muntjac deer were found in the car park of the South Bucks Star offices, at High Wycombe. Was this a grisly prank? Well, judging by the state of the unfortunate victim, some large predator had feasted upon it over night and left the grisly yet clean remains behind.

A Trevor Smith examined the carcass and told the press that: *"This is not the work of foxes. The animal's rib cage has been chewed off. It is very possible that this is the work of a big cat."*

According to Thames Valley Police a puma had been sighted at Wycombe Heights Golf Course just a fortnight previous.

On the 14th October 2001 *The Sunday Mirror* briefly covered a *'Puma On The Loose'* on the Hampshire-Surrey border. According to the report locals were blaming the disappearances of domestic cats on a large cat, thought to be a puma.

"Earlier this year (2002) whilst travelling on the M26 towards Surrey, about halfway along it, I was forced to slow down by slow moving traffic. I travel along this road every day to get to Surrey where I work. The position of the slow moving traffic was unusual as it starts to build up towards Junction 5, not at this particular spot. I slowly moved along until the traffic came to the cause of the delay.

An animal had been hit and pieces of it were all over the road (all three lanes). I am used to seeing the bodies of dead animals on the roads of Kent. However, this was different. The animal seemed to have black fur. I immediately started thinking as to what it was. It was not a sheep or a fox. Not many things bother me like this did. It was a huge animal that had been hit, but I only caught a few seconds of the sight as I was forced to drive off.

Do you know any information about this incident? I came to my own conclusion that it was a black cat.

Thanks.
Richard."

Little else seems to be known about this interesting incident. Although such events are few and far between they must surely happen on occasion. Smaller cats such as lynx and jungle cat have been picked up off the roads in Kent. In Brighton during the '60s it was once reported that two or three black leopards caused a car accident as they casually strolled across a busy road one morning. However, many drivers who do hit animals such as these would probably drive off. Any cases that may be reported would probably be cleared up by the RSPCA. Why this happens I am not sure. Do authorities fear mass hysteria within the public? In fact, would a press release by the authorities concerning the death of a 'big cat' be something akin to openly admitting that these 'mythical' animals do indeed exist? Would it be such a problem if these animals were accepted as part of a rural landscape, and if so, wouldn't that mean we could monitor them more closely rather than clandestinely investigate them? Unfortunately, the lack of official research and acknowledgment of these animals is far more baffling than the actual 'big cat' mystery itself!

Large tracks of a cat were found at the Wycombe Heights Golf Course in 2002, allegedly made by the animal known as the 'Bucks Beast'. More mutilated muntjac deer were discovered in local forests and the press had a field day:

'Is That Tiger Of Yours Licensed?' asked the *Gravesend Messenger* on January 31st 2002.

'People who keep wild animals at home could face prosecution if they do not have license for them.

The RSPCA is drafting new proposals calling for greater protection for captive wild animals.

The current law protects some, but not all potentially dangerous animals such as large constrictor snakes. Most owners do not apply for licenses, and escape annual inspections.

Two years ago a poll found only thirty-four licenses in force in the south-east – far below the suspected actual number. Owners not aware of how to deal with and care for wild animals, could find them causing serious injury.'

Several reports of a large cat emerged from the Epping area, where geese were said to have been killed by a silent predator. Zoologist and friend Quentin Rose, at the time, examined a paw print found not far from Epping but concluded it was made by a dog. The print simply looked large because it had been smudged in the mud. During the 1800s a coyote was found in the Epping area also, believed to have been released from a private collection that was imported via boat.

On the night of July 20th 2002, and also August 1st, Bexley resident Stuart Campbell, had heard a strange scratching at his back door which awoke him. The man, who was a little concerned about the noises, and the safety of his three children, decided to contact the police and the RSPCA but his call was never taken serious. However, when he spotted a massive black cat in the Northumberland Heath area, his brother-in-law certainly took matters seriously because he'd also seen it. He described the animal has having shining eyes as he watched it from his bedroom window. The black intruder slinked around the garden in the darkness leaving the witnesses terrified.

Readers of the *News Shopper* in the Bexley area bombarded the news desk to report their own encounters with large, wild cats, with a majority describing a black animal, in contrast to the puma which was often reported in and around Surrey a few decades previously.

A Bernice Fogherty saw an animal leap over a fence the previous year, at her home in Avenue Road, and there were also sightings in Erith and Eltham, as well as Welling and Belvedere.

A big, black cat was seen at Plumstead Common one month after Mr Campbell's sighting. A Dave Loson watched the animal from the window of his flat as it prowled along Upton Road. He watched it for five minutes.

On the 20th September 2002 the press asked, '*Is There A Richmond Beast?*' after a large paw print was found in a garden at East Sheen by a Beverley Cooper, who contacted the *Twickenham Times*. Unfortunately, the print faded away before it could be verified. But, on the 25th, it was alleged that the Plumstead 'panther' had been caught on film by a Steve Gardiner who lives at Upton Road. According to the press this had been the sixth reported sighting in Woolwich and Bexley in just four weeks. Mr Gardiner caught the cat on film as it strolled alongside his house. It was around 7:25 am when the witness watched the animal and was also aware that his security system had picked up the animal. Although the man described the animal as being around three-feet long the image on the CCTV is nothing more than a hazy blob.

On the 8th February 2003 a black cat, the size of a Labrador dog was sighted at Belvedere. A Craig Lenney contacted me to say he'd watched the animal from just seventy-metres away, cross the road and head towards marshland at around 12:45 am. A few weeks later prints which were claimed to have been made by a puma, were found and cast in the Crayford area. The prints measured over four inches wide, and there were also sightings of a large, black animal at Princes Risborough.

The most unusual of all the reports to come from London and the surrounding areas emerged on 12th July 2003. A Nick Allen, MSc FRAS, was staying at the Tower Thistle Hotel, next to Tower Bridge when he spied a strange cat from his window. The cat, according to Nick, was mooching around the taxi entrance to the hotel from the north end of London Bridge. He described the cat as, *"...over large, russet colour and had a brush on the end of its tail. It had very large and erect ears. I assumed it was an urban fox – but it wasn't – it was feline, far too big, and not canine in its behaviour."*

"What was odd," he added, *"is that someone else thought it odd too. As it cowered in the undergrowth, a passing person noticed it and approached it. The animal ran off – and I saw the last of it as it ran away below past the restaurant window."*

What was this extraordinary cat? Well, there is the possibility Mr Allen had seen a lynx, as he never mentioned the tail being overly long, the brush could well have been the bobbed tail the lynx is known for, as well as the erect ears. The caracal (*Felis caracal*), which are sometimes confused with the lynx, also have large, erect ears and are rusted in colour, although the tail is longer and more slim. These beautiful felids are native to Africa and parts of Asia and can reach up to around three-feet in length, whilst another contender is the jungle cat (*Felis chaus*), also known as the 'swamp' or 'reed' cat, native to Africa and Asia. The jungle cat, lynx and caracal could quite happily live in the U.K., the caracal and jungle cat, in their countries of origin live on the fringes of human settlement, feeding off rats, birds and mice and can remain in these built up areas without detection. The jungle cat is also able to breed with the domestic cat and there have been many reports of animals resembling this cat across Britain, and this also makes for interesting hybrids. However, reports of the caracal are less common. Also, because these animals are smaller, monitoring sightings of them is nigh on impossible, which is why the animal seen near Tower Bridge has remained so elusive. And, until it is seen again, we'll never be able to put a true identity to it.

At the beginning of April, and certainly not a Fool's Day prank, the Bexley 'big cat' reared its head again. This time it was sixteen-year old Daniel Brown in Welling who disturbed the creature as he let his own pet cat in through the back door. Daniel boldly shouted at the big, black cat that was squatting at the end of the garden and it leapt out of sight.

A year later residents of Barnehurst, in particular eighteen-year old Daniel Monck, and his sister Tracey, of Cheviot Close, were concerned about the animal. Rather oddly, the *News Shopper* said that Mr Monck described the animal as something akin to a large dog mixed with a sheep! The witness stated, "It's too big to be an ordinary cat, it has a huge head and big paws and pointed ears."

The grandmother of Daniel, eighty-four year old Nell Hawes also saw the animal and feared for her life around the time, and refused to go outside in case the creature was prowling around. Several neighbours reported that a large cat had been seen mooching around the dustbins and feeding off scraps.

On the 20th of April a female witness, first name of Heather, saw a black leopard from her window. It was around 6:20 pm when something caught her eye rummaging around the undergrowth near the rail track nearby. The animal eventually slinked out of sight. The woman told me, *"I couldn't believe it when I saw the big cat. It wasn't the first time it had been seen in the area. My sister had seen it when she was visiting me about a year ago."*

There are many sightings of large cats along railway lines, and there are several reasons for this. These animals use the lines for navigation, and many railway lines provide enough cover, especially at night, for a large cat to move from A to B of its territory. A lot of the sightings in built up areas such as Bexley can be explained by the railroad theory. These lines often run along the back of gardens, so any large cat could lay up in a backyard over night after it has spent the evening hunting along the tracks. Also, there are many tracks in Britain which are not often used, or maybe disused. Again, these animals can travel along these, and often remain out of view. We must also realise that the country is gradually being carved up by new land structures. Rail links are slicing their way through the countryside to enable easier ways of commuting, and many sightings do involve commuters who are gazing from their window admiring the countryside. The abundance of food along railway lines is also a major factor as to why these large cats roam there. Foxes, mice, rats, and rabbits provide a mini restaurant for the felids which stalk the rails.

During early summer of 2004 several sightings emerged from the Fawkham area, of a large black cat. One motorist observed a well built melanistic leopard holding a rabbit in its mouth and on Tuesday 29th June Mr Barton, driving on the M25 from Dartford to Swanley saw a large black animal walking across open fields towards a motor cross area. The one detail that caught Mr Barton's eye was the length of the animals tail. At 11:00 pm on the 12th July 2004 a witness encountered a puma in Bexley. The eye-witness made a statement:

"I came out of an alley from Heath Road leading to Cold Blow Crescent and was faced with something sitting next to my car. At first I thought it was a fox, but when I approached it got up to move on and it looked like a big cat. It walked off up my neighbours drive, stopped and turned its head to look at me. When it saw I was still coming (I was walking up my footpath to my front door), it went to the next drive and again turned and stared. I looked at it for a good five minutes without either of us moving.

It definitely went up the drive it was standing on and through the gate which had been left open. I see a lot of foxes in Bexley and this definitely was not a fox. It didn't walk like a fox, it walked in a feline manner and was a sandy kind of colour – possibly a bit darker on the back. It was approximately three-foot long, not including the tail and when it was standing next to my car, it's head was about level with the bottom of the window. It's face was quite wide. I didn't notice spots or stripes on it, but it was a little dark. I've had a quick look on the internet and it looks a bit like a lynx, especially the face, or it was something very similar."

This excellent description seems to describe a puma as there was no mention of a shorter tail which the lynx has.

A Leatherhead resident claimed to have photographed a large brown cat which the press connected to sightings of the Surrey puma. A Bruce Burgess, aged sixty, was surveying sites for an artificial lake, around Brook Willow Farm, Woodlands Road, when he snapped something in the distance near low branches. However, like so many other photographs of alleged big cats, this one also proved to be too fuzzy to determine what exactly was in the frame.

During October of 2004, albeit at Halstead, in Sevenoaks, a woman found her pet cat decapitated and bereft of its tail. The poor victim also had a large bite mark on its back. The shocked woman found her cat in her courtyard on the 26th at 8:00 am. Such a find merely added fuel to the fire that something more than a sinister human was around slicing and dicing moggies. Many fingers pointed to the so-called Otford 'panther', a large black cat seen for the last twenty or so years around Sevenoaks, although sightings date back almost a century.

The spring of 2005 kicked off with one of the most remarkable big cat stories ever to grace the pages of any newspaper. An Anthony Holder claimed, that he was attacked by a big, black cat during the early hours of 22nd March, whilst in his garden at Sydenham, a few miles from Penge. The witness claimed he was looking for his crying kitten at the bottom of his garden, which backs onto a patch of woodland, when the six-foot long predator leapt upon him. He told the press, *"All of a sudden I see this big black thing pouncing at me, knocked me flying. I just didn't know where I was, and the next thing there was this big black figure laying on my chest."*

Mr Holder, a father of three, allegedly suffered cuts and bruises after his encounter and was treated by paramedics. After the incident he ran indoors and phoned the police but when they arrived at the scene there was no sign of the mystery attacker. Tony added, *"I could see these huge teeth and the whites of its eyes just inches from my face. It was snarling and growling and I really believe it was trying to do some serious damage. I tried to get it off but I couldn't move it, it was heavier than me. I was scared. I really thought my life was in danger but all I was worried about was my family. It was an absolute nightmare."*

Ashleigh, the fourteen-year old daughter of Mr Holder, claimed she had seen her father being attacked by the animal, and a local painter and decorator, Billy Rich, believed he saw the animal that afternoon and a handful of other witnesses came forward to report possible sightings.

The Daily Mirror reported that the attacker sank its claws into Mr Holder's fingers and also swiped at his face, although no marks consistent with that of a Leopard attack seemed evident.

As per usual a degree of hysteria boiled in the local community, with Sydenham Girls' School closing, and police warning residents to stay indoors. *The Daily Mail* of Wednesday 23rd March 2005 claimed that:

'As Mr Holder was being treated by an ambulance crew in the street, he says he saw the beast again. "It was strolling past the back of the ambulance as if it didn't have a care in the world." said Tony.

Reports from the time seem inconsistent and several statements seem confused and the evidence does not add up to the possibility of an attack by a large cat. Whilst a cat, such as a leopard may feel threatened if cornered, or if possible prey, such as Mr Holder's kitten, was being taken away, I do not see any reason to suggest that the witness was in fact attacked. Sightings have come from Sydenham and the surrounding towns, but the injuries Mr Holder sustained are not consistent with marks that would be left by a leopard attack. Injuries sustained were a 'bitten' finger, one minor five-inch scratch to the face and a cut wrist. Such an animal has extremely large and sharp claws, which would lacerate deeply even if the swipe was merely to threaten rather than kill. Also, Mr Holder mentioned seeing the, "…whites of the animals eyes."

This is not consistent with the eye colouration of a leopard, which is more green-yellow when reflected in light, unless of course he was merely being dramatic. Even more dramatically, the police arrived on the scene in the form of an armed response unit armed with taser stun-guns, but failed to find the animal. We can compare such an incident with the case the South Harrow puma, which was owned by a local man in 1974 who often walked his prized pet through the local streets. However, on one occasion, during the November of '74, things got out of control. The owner in question casually strolled into the *Farm House Pub* with his puma on a lead. After a short while several locals began to feel uncomfortable in the presence of the wild animal and so the man was asked to leave. Although he complied with the requests of the staff and customers, the beast didn't, and in turn went berserk. The landlady at the time commented that, "It took the man fifteen minutes to get the puma out of the pub and into his car, during which it tore off the man's glove and ripped open his hand", this was after it had caused severe damage to the chair upholstery in the public house, as well as damaging tables, smashing glasses and demolishing the bar in its frenzy. Then, the cat decided to shred the car seats and the police were called to the scene where, after a short time, they towed both the vehicle and the aggravated felid away. Later, the man was charged with being drunk and incapable. But what happened to his cat?

In 1975 MP Peter Templemore claimed that:

"Someone sooner or later will get killed", in response to a strange incident at Acton in London where an estranged husband dumped his puma in the back garden of his former home, with wife and kids trapped inside screaming for help. The man left a note saying that he had nowhere else to put the cat, and it took the local police and RSPCA two hours to get the terrified family of the man out of the house. So, if Mr Holder really was attacked by a cat capable of shredding wood, smashing glasses, tearing car seat leather etc, etc, where were his injuries? Compare this also to cougar attacks in America, where people have been killed, and the same can be said for victims in Africa with reference to leopard attacks.

On the 21st March 2005, a woman also saw a black leopard, this time in Sidcup. She was putting some washing out at around lunchtime when her garden fence began moving and she saw something black leap up. She phoned the RSPCA who took her details but never seemed to follow the report up.

In 2002, a Gravesend man claimed he was slashed across the hand by a lynx after he disturbed it near his home. The cat was carrying a rabbit in its mouth and witness, a Mr Cole, attempted to approach the animal, which at first he took to be a fox, in turn receiving a nasty set of scratch marks across his hand. Even this encounter at the time

In less enlightened days, exotic cats were often kept as pets

was treated with scepticism by some researchers, but the injuries Mr Cole sustained were far more horrific and in common with a 'big cat' attack than Mr Holders. Mr Cole never blamed the attack on the animal, but instead stated he pretty much deserved it, the animal was merely retaliating after it felt threatened.

Borehamwood had a minor cat-flap during early April '05 when several animal remains were found in fields near Berwick Road. The carcass of a swan was found by a six-year old boy and his mother also claimed that various other animals had been discovered, half-eaten, as well as large paw prints. Chickens, ducks and foxes had allegedly been slain in the vicinity sparking rumours that a leopard or puma had been prowling around the neighbourhood. At the time sceptics claimed that the swan death may have been down to the power lines in the area, and that the poor bird had struck one of these whilst in flight. However, the other grisly remains discovered in the area seemed to suggest some kind of killing machine was haunting the undergrowth.

Meanwhile, at Eltham around the same time, a devoured fox was found in the back garden of a Laura Downes, who lives at Westmount Road. When she first went into her garden she found a dead fox which she didn't think was anything unusual but one hour later when she returned, the fox had been stripped clean. The following day another eaten fox turned up, this time covered in maggots, suggesting this carcass was older. The terrified woman called the local Wildlife Officer who stated, *"It was horrendous. Whatever did that to the first fox did a good job. I've never seen anything like it. I was scared."*

A female witness named Vicky saw a melanistic leopard in Ewell, Surrey on 16th April 2005. She and her housemate were staring out of their kitchen window when something black caught their eye down the bottom of the garden. The animal had a feline gait and was as big as a medium sized dog. The cat then disappeared into the neighbours garden.

In May, one of the Surrey felids was caught on film at Winkworth Arboretum by a Harry Fowler of Guilford. He shot the footage near Godalming two weeks after there was a report of a similar cat in a tree at Whitmoor Common. The cat on the film was pointed out to Mr Fowler by a female witness, who spotted it near the boathouse at

Phillmore Lake. The cat in the video is a rusty brown colour and roughly the height of an Alsatian dog. Aurrey Wildlife Trust manager believed the animal in the film was a lynx although it could also be a puma.

Shortly after the video footage was taken of the mystery cat, many Surrey residents called the press to report their sightings, proof then that the Surrey puma and friends were still healthy, despite a decrease in Surrey sightings since the original flap. These current populations suggest that twenty to thirty years ago large cats such as the puma were indeed breeding, and there is also the small possibility that some cats are still being let go into the countryside to add to the already growing number.

Winchester's Marwell Zoo liaison officer Bill Hall commented, *"Personally I believe there maybe a big cat somewhere but not leopards, and definitely not black ones."* A very strange, almost amateurish view from a zoo officer! He continued however that, *"...I am quietly confident that a puma could survive."*

He then went on say, *"We get calls from people saying a big black animal jumped in front of the lights of their car, what do we think it was, and I say it was probably a Labrador dog."* In my opinion, such a statement is a joke. Most witnesses know when they've seen a big black leopard, and I can't see why there would be an abundant population of stray, yet very uncharacteristically elusive black Labrador's roaming the countryside. Mr Hall accepts that the puma could survive in Britain, but not the leopard despite the fact that menageries during the Victorian period and also private collections housed both species of cat in equal measure, during the 1960s and previous, and even up until today.

The 'beast' of Bexley was seen again in the May of '05 by care worker Jim Hornby who works near Bexleyheath Broadway. At around 11:00 pm one night, whilst talking to his supervisor, a something black caught his eye. A cat-like animal was creeping along on its belly near to Jim's car, as if it was stalking. The cat also seems to have left a calling card, a huge paw print which Mr Hornby discovered in a sandy area in the garden. The print measured five inches across.

I was contacted by nursing officer Nicole Webb regarding the sighting and the details she gave me were far more in-depth. She told me the following, "Whilst on duty during May '05 I was stood talking to staff by an open door and saw what looked like a man on his hands and knees. We get a few intruders at the site, so myself and Jim went outside to tackle him. Jim was side on and I was behind what we quickly realised wasn't human. Our bowels emptied quicker than an enema could manage! All I saw was the back end, haunches and hockey-stick tail. A six-foot fence was cleared and it was gone. Jim's view was no better. He is a regular visitor, and we've heard him crashing out of a tree onto the roof of the minibus, leaving a big dent in the roof. We've witnessed the trees shaking violently on the most still of nights."

'Comment – Is The Surrey Puma On The Prowl Again?'

'Tales of unidentified animals roaming parts of Britain are not unusual - definitive proof of the existence of the Beast of Bodmin Moor and the Loch Ness Monster has been sought for years.

While not having an enormous international profile, tales of the Surrey puma have become a local legend.

Our first recorded instance of a big cat sighting was more than 40 years ago when a Munstead workman, George Wisdom, saw a golden brown animal of around three to five feet in length while he was blackberrying one lunchtime. Plaster casts taken of the paw prints were said to be those of a 'large member of the cat family'.

Sightings continued throughout the sixties, prompting headlines such as, "puma could solve deer problem", "A plane may hunt the puma", "Green-keeper meets puma on the second tee" and the brilliant, "I trod on puma's tail - and hit its nose".

The story appeared to capture the imagination of Advertiser readers, as the many letters in its archives show. Photos of the beast started to appear as well, some blatantly cats, others barely visible.

Sightings in the 70s were few and far between compared to the preceding decade, and the paper even received this letter: "Sir-I am feeling horribly neglected; couldn't someone see me again soon? -Yours, etc. The Surrey Puma".

The 80s were a non-event for the big cat, just like it was for the rest of us, and only a handful of stories made it into the paper.

The 90s to marked a glorious return, and pictures and sightings dotted the decade. After a short lull the beast has returned, and the myth lives on.

One question though. No-one has claimed there is a breeding pair, how long do these cats live?'

-*Surrey Advertiser* 23rd MAY 2005

On 21st June 2005 the Bexley News Shopper reported another cat sighting, this time at CorralineWalk, Thamesmead. A Belinda Bull was looking out of her living room window at around 9:30 pm, the previous week, when she spotted a large animal around some dustbins. Although she described the animal as having a massive head, big paws and a long tail, no description of colour was given.

A month after the sighting a fifteen-year old male from Farnham Road, Welling, saw a large cat whilst on his way home from a friends house. It was around 10:00 pm on July

12th when he heard a rustling sound in the nearby bushes. As he strolled by he glimpsed a black cat in the undergrowth behind a metal fence. The witness called the police who followed up the report.

This Is Local London asked, '*Is A Panther Loose In Harpenden*' after a sighting of a large cat on October 3rd 2005. A woman walking her dog on Bower Heath Lane reported her sighting of, what the website claimed was a puma, to the police. A security firm in Hertfordshire believed they had caught the animal on CCTV in September although what became of this footage I do not know.

2006 was only eleven days old when the press reported an early January encounter involving dog-walker John Costin spotted a large cat at Churchfield Woods, in Bexley Village. The man was with his dog 'Mickey' when he observed an animal in long grass not far away. The cat, which he never got a clear look at, seemed to have a tortoise-shell pattern to its coat and measured only around two feet at the most, from nose to rear. John never got a look at its tail but the animal seemed very wild, and he believed it must have been a lynx or a bobcat, which is slightly stockier than the lynx with a rusted coat. However, the report of the animal didn't seem to be consistent with any other cat sightings, so the jury is still out on the matter.

One month after the strange cat sighting at Churchfield Woods, a *News Shopper* reader, a Ms Cashnella, came forward to file a report of an animal which she saw some time during 2004. Her interesting letter read as follows:

'I too was a non-believer until I saw this cat for myself. Around two-and-a-half years ago I had a mare who was ready for foal and I went to check on her about 11:30 pm.

My horses are kept on Southmere Park, or, as some might know it, Erith Marshes.

There it was, stalking through the reed bed. It was about two-feet in length and about eighteen inches high. I searched the internet to see if I could find out what I had seen and there it was, a jaguarundi.

These cats come from South America and live in habitats ranging from semi-acrid scrubland to swamps.

Its main prey is birds, rodents, rabbits and reptiles. Hence there is no savaging of domestic or wild animals or humans.

These cats can range from black to pale grey-brown or red in colour. They are about thirty-three inches long and weigh around ten to twenty pounds.'

This sighting was extremely intriguing, and certainly could have explained some reports of unusual cats around London. In fact, there are many eye witness reports nationwide which often remain inconsistent with the usual leopard, puma and lynx sightings, and this is because there are other, smaller species out there. Caracal, ocelot, and jaguarundi reports do exist. Some of these cats may be individuals that have escaped from private collections or zoo parks, and a majority of times they are sighted, witnesses may well not know exactly what they have seen, unless they are quick to act like Ms Cashnella, who identified her marshland stalker via the internet.

Britain is perfect habitat for a majority of large cats, and the smaller cats are even less likely to be seen. In fact, animals such as lynx can remain undetected for many years. We just do not know what cats are roaming our back yards at night, the variety really is surprising.

The jaguarundi however, despite the Erith Marshes report, most certainly isn't the cat behind most of the sightings, because quite clearly it isn't a big enough animal. So, once again we go back to the puma and the melanistic leopard.

On the 15th February *This Is Local London* claimed that the elusive 'beast' of Bexley had finally been photographed. Debbie Marshall of Crombie Road, Sidcup, snapped an animal whilst she was looking out of her bedroom window the weekend previous. A black animal laying in the grass caught her eye so she reached for her digital camera and took a photo as the animal mooched around the field.

Unfortunately, despite a good description of the animal, the photograph is once again dubious, although most certainly shows a black object in the long grass quite a distance away, which Mrs Marshall estimated at around one-hundred feet. The photograph, which appeared in numerous newspapers over the course of a few days, appears to show the head and big shoulders of a black leopard skulking beyond the trees. This seems promising, although some sceptics argued that the image could be anything from a domestic cat to a black sack. Mrs Marshall claimed that her neighbour's pet cat was also in the field, yet looked around seven times smaller than the animal she'd photographed. However, despite the excitement of the witnesses, which included Mrs Marshall's three children, the photo has once again been filed alongside so many other 'big cat' pictures labelled under the 'maybe' category.

On May 30th 2006, *IC South London.co.uk* reported on the '*Man To Keep Leopards In Back Garden*', with an exclusive report by Richard Porritt. The story concerned Mr Todd Dalton, known locally as the Leopard Man of Peckham, due to his mini leopard sanctuary which he'd constructed in his back garden. However, the man faced much fear and loathing from neighbours who believed that his animals were threat to youngsters and so a ban was bestowed upon him, preventing him from keeping such animals. Mr Dalton however paid up to twenty-five thousands pounds to contest the decision and in May his ban was overturned after he dragged Southwark council through the courts.

Reporter Richard Porritt stated: *'After the hearing at Tower Bridge Magistrates' Court, Todd told the South London press he was, "...delighted", but the action had left him broke.*

"This appeal has cost me twenty-five thousand pounds and I am not applying for costs because I do not think the taxpayers should have to pay", said the internet entrepreneur.'

According to the website, Mr Dalton had built large cages in his garden to house the leopards but councillors refused him permission to keep the animals despite the go-ahead from police and vets and also working in compliance with the Dangerous Wild Animals Act.

On the 11th June 2006 a Clara Story, for *This Is Local London* reported that another domestic cat had been found without its tail, this time in New Maiden, south-west London. At the same time, I received several reports from rural Maidstone of many domestic cats missing. One woman found the head of her pet cat on her patio one morning, a grisly find which befell another Maidstone resident some four years previous. Both witnesses had heard deep growling and a heavy rustling in the bushes the previous night.

The attack on the domestic cat in New Maiden was once again blamed on a knife-wielding psychopath due to the cleanness of the wound. The poor cat probably died of a heart attack.

So, large sandy-coloured cats roaming Bexley, black 'panthers' slinking through back gardens of Sydenham, and large eared felids mooching through the capital. Just how did they get there? Well, the most common theory, which you'll read much about in this book, does concern exotics purchased during the 1900s which may have been released during the 1960s and '70s, and after. Could cats obtained and then released after the introduction of the Dangerous Wild Animals Act be responsible for the populations of today? Partly, but not fully.

This is no modern mystery, but the fact that you could buy a large, exotic cat from a London store such as Harrods, not too many moons ago does lay the blame to some extent on exotic purchases during a period when your average novelty pet usually came in the form of a slinking, shiny black leopard via the local pet shop or trader newspaper. To some, these animals would have made great fashion accessories, whilst to others these animals would have been great guard dogs, and maybe even sweet pets, but there would have been those who could not have handled such cats, and certainly when the 1976 Act was passed, the license fees would also have proven too hard to swallow and so a few cats must have been released into the wilds. There's no way that every cat purchased would have been found a home in a local zoo and I'm pretty convinced that a number of zoos would have turned quite a few owners away, maybe hundreds, as they flocked to the gates to bid farewell to their beloved pets. There is also the possibility, and as shown by the capture of the Barnet lynx, that wild cats are still easily obtainable today. Whilst those which are kept legally are monitored by local councils there are still those people out there intent on keeping exotic animals in makeshift pens which are not always adequate enough to house an animal for that long. I'm aware that very recently two female puma's and a lynx were released locally into an area I will not mention here.

However, the negative side to this theory is that for the animals which roam today to be offspring of animals let go in the '60s, there'd surely be more sightings? Well, in reality there are hundreds of sightings which each year make their way to the press, the police, local researchers and zoo parks, and there must be almost as many sightings which go unreported. However, sightings of normal leopards hardly exist, so this factor could point to the swinging 1960s fashion-trend of having the customary black leopard at one's side, as well as the occasional puma and other smaller cats. There may also have been a few people who owned cats that weren't quite as knowledgeable and were unsure as to what they were looking after. A cat such as the golden cat can also have a dark coat, but someone may well have purchased one thinking it was in fact a black leopard, and we can see by many modern reports that so many people aren't sure of the difference between a puma and a 'panther', so who knows what kind of cats were being obtained.

For me though, the strongest theories to explain as to why these animals exist today concerns the Romans and also travelling menageries, for how else can we explain these brief reports of a Surrey mystery animal and others, as reported in Charles Fort's important 1931 book *Lo!* in which he writes, *"In the Daily Mail, March 19, is an account of an extraordinary killing of sheep, 'by dogs' near Guildford (Surrey)...fifty-one sheep were killed in one night. A woman in a field – something grabbed her. At first the story was of a marauding panther that must have escaped from a menagerie."*

There is also rumour of a large, black beast from AD 940, which was said to have haunted Flixton in Yorkshire. Some claimed the creature was conjured during a dark ritual, and its gory rampage included humans as well as livestock and dogs. In 1455 a Percival Cresacre was killed by a mysterious felid at St Peter's Church in Bamburgh, South Yorkshire. The unfortunate victim was riding horseback through the woods when a ferocious cat leapt on him from a tree and clawed him to death. The horse took off leaving the man to fend for himself but the cat clung to his back, injuring him severely. The man was found next day, weak and bloody, by the church porch and died later. The Cresacre family, it is said, have a tiger-like cat on the family arms and also on the tomb. Some have suggested the felid is a wildcat, or 'wood-cat' as the animal was called at the time, but it seems unlikely that the wildcat, resident of Scotland now, would have weakened a man to the extent of death. It seems more likely that Mr. Cresacre was attacked by a leopard, which was laying up in a tree.

Gamekeepers and hunters killed the English wildcat prior to World War I, but in Scotland the animal is abundant. In his superb book *Mystery Cats Of The World*, Karl Shuker describes the felid as follows, "The background colour of a pure-bred British specimen's pelage is buff-grey marked with several vertical stripes running down from the dark dorsal stripe to the belly. The limbs are also traversely striped, and the black-tipped tail is encircled by a number of rings (of which at least the last two or three are complete). Its forehead bears four or five longitudinal stripes running down to the neck's nap, where they converge to form the dorsal line. The wildcat can be distinguished from most (but not all) domestic tabbies by its absence of blotched markings. Generally, the wildcat is about a third larger in overall size than a domestic, its head broader, its teeth larger, its body stouter, its limbs longer and its tail shorter." The wildcat has never been indigenous to Ireland but sightings occur, as well as in England still, many years after it was allegedly wiped out.

The legend of Black Annis may also play an important part regarding sightings of early mystery cats. This particular fiendish fiend was considered supernatural at the time, during the eighteenth century, but the habits it displayed seem now to point to a black leopard. Black Annis was said to have prowled the Dane Hills of Leicestershire, and was born from an obscure poem which also concerned an old hag character who inhabited a dank cave. Annis was said to have leapt onto victims whilst hiding in trees and sink teeth and claws into the flesh of prey. Indeed, from such poems and legends we can see how such animals have faded into myth and the realm of ghosts.

In most cases, a large cat such as a leopard would only take one sheep to eat over the course of a couple of days, administering a throat bite to bring the animal down, and at times dislocating the neck. No other animal in the British Isles kills like a leopard or a puma, but there are enough rabbits and other small prey so an animal such as a leopard will not often expel energy on a large ewe unless very necessary. These animals are stealthy hunters that devour the innards of large animals such as sheep. Prey will often be stripped and stashed high up in trees, numerous reports from Bodmin during 1982 and involving eaten sheep described how the flanks and shoulders of the victim had been torn away and the flesh rasped clean. However, even if a cat had young with it, it seems unlikely that even to train its offspring, that the 'family' would injure let alone kill so many of a flock. Dogs are certainly the most likely candidates for such attacks as they nip and tear aggressively and randomly, and not to eat, but out of sheer spite. However, some time during the 1920s a mystery killer was attacking sheep and goats in Inverness-shire in Scotland. During the flap of attacks, where many prints were found in the heavy bogs, farmers described the predator as large, fierce and yellow in colour. One strange attacker was shot at, and described as a large cat. Another, also a large cat was caught in a trap. These animals were identified by London Zoo as lynx.

Over 100 years previously a creature that become known as the 'Girt Dog of Ennerdale', at Cumberland was terrorising sheep on the England/Scotland border. The critter was killing all prey with a throat bite and draining the carcass of blood. A dog that was shot in the area convinced few that the killings would stop, for they believed that something far larger and more efficient was at large, like a lion. No-one at the time ever reported a barking creature, but something more akin to a felid. Similar sheep kills were also occurring in parts of Ireland around the 1870s, at Gravesend in Kent in 1905 and in Gloucestershire during the same year. The press identified this particular animal as a jackal. In the following year the *Daily Mail* reported that some vicious animal sighted near Windsor Castle, in Berkshire, had bitten almost a dozen of the King's sheep, all of which had to be destroyed.

On October 14th 1925 the *Daily Express* reported on a big, black creature stalking the night in the district of Edale, Derbyshire. This silent hunter left a trail of corpses in its bloody wake, much of these having heads and legs ripped off. Despite several hunts for the killer, nothing was ever flushed out of the undergrowth. The same could also be said for an unidentified back-breaking phantom which haunted Llanelli, in Wales during 1919. This particular prowler killed many rabbits but never ate them. It simply left piles of poor, broken bunnies. Meanwhile, in 1939 two ghost hunters visited the eerie Borley Rectory in Essex which at the time was in ruin due to a bad fire. When they entered a room upstairs they were confronted by a big, black creature squatting in the corner that gave off an evil menace. One of the witnesses, author James Wentworth Day, drew a weapon to shoot the beast but was prevented by his colleague. Then, when the pair strolled through the orchard, a large, dark cat-like animal sped passed them into the black of night.

So, where did these mystery maulers come from? Well, the Romans invaded Britain around 2,000 years ago. The scimitar cat became extinct more than 500,000 years ago, the cave lion died out in the United Kingdom some 50,000 years ago, and the last leopards to inhabit Britain died out some 12,000 years ago during the last Ice Age, so such beasts most certainly are not survivors from then. The puma meanwhile has never been a species indigenous to Britain. Whilst Great Britain once sustained a variety of species of felid, only the wildcat, confined to Scotland, survives.

A basic timeline of events shows that Julius Caesar headed the first invasion in 55 BC, but withdrew his advances. A century later the Romans invaded again in AD 43 under the reign of Claudius, when they landed at Richborough, in Kent, where the remains of an amphitheatre exist. Before their invasion areas such as London did not exist. It was they who founded it, in AD 50, calling it Londinium. An amphitheatre was discovered in the city in 1988 and is now open to the public. Other amphitheatre remains can be found at Chester, Colchester and Gloucestershire, namely Cirencester, formerly the Roman town of Corinium, which is Roman Britain's largest amphitheatre. The Romans conquered Wales, there is an amphitheatre at Caerleon, Gwent, and the north in AD 70 and Scotland in AD 140, where an amphitheatre is situated at Inveresk, Edinburgh. They did not withdraw from Britain until AD 401. During this time, however, they had the amphitheatre, their centre of entertainment; a bit like television is for much of today's general public on a Saturday evening!

In these amphitheatre's, Roman citizens would flock to watch gladiators fight wild animals such as bears, lions and leopards. Often, the fighters who took on the beasts were slaves or criminals being punished for their crimes. In many cases, these gladiators would die at the paws of an animal such as a big cat. Such a spectacle may well have influenced the more modern travelling menageries, where strange and exotic species were collated from remote regions in order to attract the curious public.

The sheer gore of a Roman gladiator battling it out with a starving leopard would have provided entertainment to thousands of roaring citizens. The volume of animals used in such battles is quite incredible and many would have been slaughtered in the most barbaric of fashion.

The *Endangered Species Handbook* writes:

"The tradition of killing animals for pleasure has a long history in Asia and Europe. So popular was hunting in ancient Rome that mosaics and paintings often depicted this pastime as a heroic activity. Slaughtering animals was considered a form of entertainment, and people scoured the countryside for bears, lions, stags and boars to pursue with spears and dogs (Attenborough 1987). As the Roman Empire grew to encompass the entire Mediterranean basin, its citizens travelled throughout the region to hunt and bring back animals to be killed in primitive contests in the coliseums of Rome and other cities. The coliseum games continued for more than 400 years in more than 70 amphitheatres, the largest seating up to 50,000 people on stone benches arranged around a central arena (Attenborough 1987).

Roman emperors carried favour with the public by upstaging their predecessors in killing more animals and producing more spectacular displays of slaughter (Morris 1990). Emperor Titus inaugurated the Roman Coliseum by declaring 100 days of celebration, during which enormous numbers of animals were speared by gladiators. On the opening day, 5,000 animals were slaughtered, and over the next two days, 3,000 more were killed (Morris 1990). The caged animals were kept underground in dungeons where they were not fed, and on the day of the festival, they were hauled in their cages onto lifts that brought them into the centre of the arena. As the crowd roared with excitement, drums were beaten, trumpets blown, and the terrified animals were set loose (Attenborough 1987). Sometimes the animals were goaded to attack one another, and at other times, men armed with spears and tridents pursued them around barriers made from shrubs in imitation of hunts in the wild (Attenborough 1987). One arena hunt resulted in the killing of 300 Ostriches and 200 Alpine Chamois (Morris 1990).

Lions, tigers, bears, bulls, leopards, giraffes and deer died after being tormented, stabbed and gored (Morris 1990). Big cats that had been starved were released into the ring where a human slave or prisoner of war was lashed to a post; the animals clawed at the person before they themselves were speared and stabbed by gladiators (Attenborough 1987). In some of the larger slaughters, 500 lions, more than 400 leopards, or 100 bears would be killed in a single day (Morris 1990). Hippos, even rhinoceroses and crocodiles, were brought into these arenas, and sometimes gladiators employed bizarre methods of killing such as decapitating fleeing ostriches with crescent-shaped arrows (Morris 1990).

The Roman audiences cheered these brutal slaughters enthusiastically as a rule, but when 20 elephants were pitted against heavily armed warriors, the screaming of these gentle animals as they were wounded caused the crowd to boo the emperor for his cruelty (Morris 1990). This did not stop their use in the games however. These slaughters virtually eliminated large mammals from the Mediterranean area."

Many animals were imported to Britain for the pleasure of the grisly arena battles. This shipping industry was a huge operation, with some animals being obtained from the west of India. Such wild animals were considered to be special gifts which many a barbarian monarch would offer to his overlord. In Sicily, close to the village of Armerina there exists a fresco – mural painting - showing trapping devices used to capture and crate wild animals for export. Animals such as leopards would also be caught in nets dropped from trees, and third century poet Oppian even spoke of leopards being drugged by strong wine leaked into waterholes to enable capture.

By the fourth century the attraction of the amphitheatre died out, after the adoption of Christianity which became the official religion of the Roman empire. The long production line of imported animals also dried up, and the public gradually became repulsed by the bloody battles.

If, as the facts state, the Romans imported large cats in such abundance, what are the possibilities of escapees? Very likely I would say. And who would know, or even care at the time ? No-one. So then there we have our first piece of the jigsaw. Leopards and maybe a variety of smaller cats would have been imported to Britain, some, maybe more than a few, would have escaped, and if a majority of these were leopards, as the history tells us, then already a population would have been present, a seeding vital enough to produce a breeding population of large cats.

There have also been rumours of airmen, especially during World War II, bringing large cats such as puma, to these shores as mascots. Researchers claim that some of these animals may have been let go over here, but even so, would these kind of cases cause an abundant population of felids? Well, although the answer to that is no, the animals which may have been released would simply have added to an already growing, and healthy number of wild cats already prowling the countryside of Britain. The same could also be said for the smaller cats sometimes used on boats to catch rats and mice. The ocelot and jungle cat are typical of the smaller species of felid used for such a cleaning operation which would have taken place on many boats and ships. Such animals may have been let go, or escaped once the boat had docked at its destination.

It is the travelling menagerie which provides another important clue with regards to how many of the cats of today got here. The origins of the travelling menagerie lies with the Romans and also Royalty in Europe around the seventeenth century, who would accumulate a number of beasts for their own personal enjoyment, although the first ever recorded zoos dated back to 2500 BC. The first royal menagerie was held at Woodstock Manor, know known as Blenheim Palace, during the reign of King Henry I, 1106-1135. Three leopards were given to King Henry III at Woodstock by Roman Emperor Frederick II, which were moved to the Tower Of London in 1235. The Zoological Society of London was formed from this collection and what is now known as the Regent's Park site was formed in 1830. On the 24th October 2005 the *BBC News* website reported:

"Two lion skulls unearthed at the Tower of London have been dated to Medieval times, shedding light on the lost institution of the Royal Menagerie.

It also shows the relationship between England's early monarchs and the 'king of beasts' was not just a symbolic one. The lions may have been among the first to turn up in Northern Europe since the big cats went extinct in the region at the end of the last Ice Age. The menagerie was a popular tourist attraction, hosting exotic animals. In addition to the lion skulls, researchers also analysed a leopard skull and the skulls of nineteen dogs. The best preserved lion skull was radiocarbon dated to between AD 1280 and 1385, making it the earliest Medieval big cat known in Britain. The leopard skull, which was badly damaged, dated to between 1440 and 1625, which covers the Plantagenet reign, the Tudors and Stuarts.

Despite their royal status, the cats were not treated with ceremony when they died, instead being dumped – unskinned – in the Tower's moat."

In the Domesday Book, Woodstock was described as, 'forest of the king', ground that Henry I used to house lions and leopards for hunting. The Anglo-Saxons were rather loose with their hunting laws as it was considered more of a sport, but the Norman kings were bound by more strict laws, their royal hunts encased with stone walls which encased the heavily forested region.

Wealthy families would also obtain exotic species, a trend which continued up until the 1970s when it was extremely extravagant to have an animal such as a black leopard in the basement. On record there are several cases of large cats escaping from their holding and being shot, the most famous incident taking place in 1530 at a spot marked as Gifford's Cross, at Chillington Hall, West Midlands, where a 'panther' escaped from its cage and was about to attack a woman and her baby before Sir John Gifford, who owned the cat, destroyed it with a bolt to the skull.

The travelling menagerie, also known as the 'Beast Show' was the next best attraction to the waxworks and theatres. Many would flock to see dancing bears, and performing lions, and in turn organisers would attempt to bring far rarer and more exotic species into the fray, such as foreign birds and extraordinary reptiles in order to compete with one another and attract a bigger audience. These exhibitions would travel the country, stopping off for a few days and then rolling, by wagon in which the animals were housed to the next destination. These shows were very much a zoo in motion, and animals were stocked overnight in dealers' yards. When the time came for the show, the wagons would be situated as to form an area and a huge, attractive façade would catch the eye of the public. By the early nineteenth century many menageries were buzzing around Britain, the most well known example being run by George Wombwell, although there must have been many obscure shows on the road also. The more popular exhibitions would not merely boast exotic animals but also magicians, lion-tamers, dancers and actors, some of which would become cult figures of their time.

George Wombwell, born in 1777 was the most famous menagerie man. It all began one day when he purchased two snakes from a man at London Docks, and began exhibiting them around the local pubs. Wombwell made a decent enough amount to realise that this kind of attraction could go down well with a bigger audience, and so began to form his own collection of wild animals, creatures which he often purchased from boats coming into London and which had been on world voyages. By 1830 'Wombwell's Royal Menagerie' was in full flow, and some fifteen wagons made up the show displayed in 1839. Within these wagons were paraded a variety of animals from elephants to llamas and zebras, but the most intriguing were the leopards, ocelots and 'panthers' which must have been pumas!

When Wombwell died in 1850 part of his menagerie was left to his niece Emma Bostock, who, with husband James, ran it from 1866 to 1884, before they moved it on to their son James who in turn sold it on to his brother Edward. It was very much a family tradition. From the 1880s the travelling show became known as 'Bostock and Wombwell's Menagerie'. By this time the rolling zoo had become a worldwide phenomenon and by the 1920s, more than seventy species of animal were put on show. The last Bostock exhibition took place in 1931 after James Bostock declared, under ill health, that no further shows would take place if there was no longer a Bostock at the helm.

A who's who of 18th and 19th Century menagerists lists over fifty names pertaining to exhibitions involving wild animals on the road. Sir Garrard Tyrwhitt-Drake, who owned the Cobtree Estate, near Maidstone in Kent, which had its own small zoo, was Mayor of Maidstone twelve times, and had his own travelling zoo which appeared at Southend, Margate, Crystal Palace and Wembley. It consisted of some ten wagons. Large cats such as leopards and puma were also exhibited at shows run by Albert Haslam - who travelled around Yorkshire and Lancashire - as well as the famous Thomas Atkins, who was a rival to Wombell's menagerie. Italian Stephanus Polito succeeded Gilbert Pidcock in 1810 to run the Exeter Exchange menagerie, at The Strand, London, renaming it the Royal Menagerie. It is said that the animals on display were housed in cages little better than travelling trucks. The Mr Pidcock is rumoured to have been the first menagerist on record, his show dating back to 1708. His collection of animals were permanently stored at London's Strand.

Thomas Stevens was a menagerie owner from 1865, in the Liverpool area. His exhibition had several leopards, 'panthers', lions and tigers. While the imposing of figure of menagerist and animal dealer John D. Hamlyn was enough to scare anyone away, his poky shop, said to have faced the walls of the London Docks, housed many small animals but in the yard there were many big cats stored. Characters of this ilk are too numerous to mention, but we can see by this procession of menagerists the vast possibili-

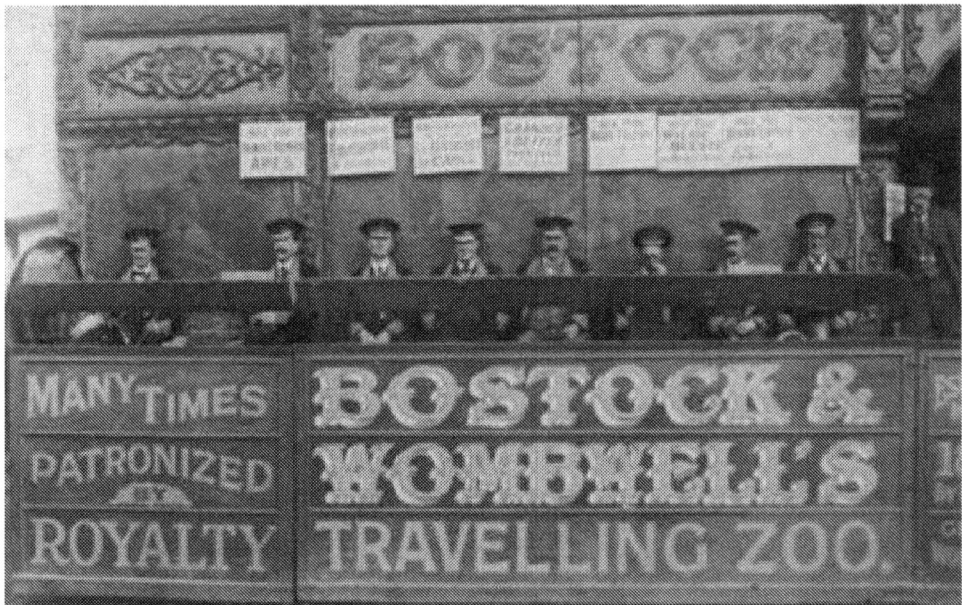

ties pertaining to populations of cats roaming the United Kingdom. Thousands of large cats must have been imported to Britain from the coming of the Romans to the travelling menageries, and escapees must have occurred on a regular basis, many of these unreported at the time.

Ballard was a menagerist known for his troupe of monkeys and dogs in 1751 which occupied Haymarket in London. One incident related to Ballard and big cat's escaping, took place on the evening of Sunday 20th October 1816, and involved The Quicksilver, a mail-coach running between Exeter and London. In 1987, the Wylye Valley Life magazine of Wiltshire, covered the story in detail, and here is author Danny Howell's take on what is now known as the 'Winterslow Lioness':

"...the coach left for Salisbury for London. On nearing the inn known as the Hut (now the Pheasant Inn) at Winterslow, about seven miles north-east of Wiltshire cathedral city, what was thought to be a large calf was seen trotting beside the horses. The steeds became nervous, and had due reason to be, for the 'calf' was in fact a lioness which had escaped from a travelling menagerie parked at the roadside. This menagerie was shortly due to appear at Salisbury Fair (the animal also appeared at the Bartholomew Fair in 1825).

The team of horses began to kick and lash out, causing the coach to sway and panicking the passengers. The lioness began leaping at the off-leader, a fine horse called Pomegranate, badly mauling him. He was, of course, fixed in the traces and could do nothing to escape the fangs and claws of his assailant. He was a former racehorse, dubbed a thief on the course but had developed such a bad temper in the stable that he had been sold to a coach proprietor, hence his second career as part of a team pulling the Exeter Mail. The guard, Joseph Pike, reached for his blunderbuss and was about to fire when the menagerie owner and his assistants, accompanied by a Newfoundland dog, came upon the scene. The owner shouted not to shoot and the dog seized the lioness by the leg, which diverted her attention and prevented further injury to the horses. In the ensuing struggle the lioness killed the dog before running under a nearby granary; a building propped up off the ground by a set of straddle stones as a precaution against rats and vermin.

The coach-driver and the guard remained transfixed on top of the coach, fearing for their lives; while the passengers, screaming at the tops of their voices, fled for safety to the inn, bolting the door behind them. An ostler employed at the inn, settled the horses, while the menagerie keepers searched in the darkness under the granary with the aid of lighted candles. The lioness, believed to be five years of age, was normally quite tame and hearing familiar voices, allowed her keepers to catch her in a sack and carry her back to one of the cages.

The owner of the menagerie was particularly enterprising because following the incident at Winterslow, he promptly purchased the wounded horse and exhibited him alongside the lioness at Salisbury Fair. This was a successful move on his part and hundreds of fair-goers paid to gaze on in horror at the horse's injuries.

The story soon attracted national attention. Not only was it reported in newspapers; the attack was illustrated by two artists, James Pollard and A. Sauerweid. The illustration by Sauerweid, although awe-inspiring, is purely theatrical. Much rarer than Pollard's, it shows the lioness attacking the leading horse with a great deal of ferocity. The passengers fly from the coach with streaming cloaks, while men with torches come to the rescue. The Newfoundland dog is, for some reason, portrayed as a mastiff. Pollard's print, is the more accurate of the two…it shows the coach drawing up in front of the inn, with the lioness plunging at the throat of the leading horse.

The story of the lioness at Winterslow has been depicted again, more recently, when it appeared on one of the five 16p stamps issued by the Post Office on 31st July 1984, commemorating the bi-centenary of the introduction to the mail coach."

So, as you can see by the examples listed in this chapter, there is enough evidence to support the current population of large cats roaming the British countryside. Add together the many menageries, private collections, inadequate facilities at shoddy zoos, animals brought to these shores at mascots during both World Wars, cats imported on boats, the Roman amphitheatre and also a few cats purchased simply to be released into the wilds on purpose, and we have a pretty healthy starting base. It also proves that the British 'big cat' situation is far from a modern mystery despite the constant reports in the press and from many researchers, that the reason these animals are here today is simply because of the last thirty years of escapees and releases into the wilds. I simply do not go for this theory, or even the suggestion that these animals have escaped from

zoos consistently enough to produce today's population.

Sightings from a handful of centuries ago may not be in abundance, but then again, there were no newspapers to record such encounters, but they certainly occurred, one of the most impressive coming from the Pilgrim's Way pathway in Kent, which runs to London, where, during the 1500s, 1600s and 1700s there were reports of strange, elusive large animals prowling the route. Legend was once born of the 'great dogg' roaming the area, as well as a monster hound of Trottiscliffe, although the details of the brief sightings seem to suggest we were dealing with a large cat such as a puma or lynx. On one occasion during the July of 1654 a rambler was even attacked and killed by a mysterious animal on the Pilgrim's Way. In 1754 a similar incident took place on the Pilgrim's Way but closer to Medway. A peddler was killed by a beast described as being a, "…lean, grey hound with prick't ears", but its behaviour, which involved stalking the man, appeared to be more cat-like. One-hundred years later in the vicinity of Burham and Boxley, a similar beast, greyish in colour was sighted by a Reverend Edward and a friend as they were strolling along when they noticed the lean animal appear several metres behind them. They described the creature as big as a calf with upstanding ears.

I'm sure that one or two cats do escape from zoos. On December 7th 2001 a press release spoke of a four-year old Indian tiger being shot after it escaped from its cage at Howlett's Wild Animal Park, in Canterbury. Fortunately, at the time of the escape the zoo was not open, but the tiger was seen to be moving to an open area and so was shot dead. However, although Howlett's has lynx, caracal and at one time a black leopard, any animals that escape would easily be accounted for, and I'm sure the same goes for many other reputable zoo parks in Britain. However, during the October of 1984 five tigers were deliberately freed from Howlett's Zoo Park. Two adult Indian tigers, Gelam and Putra, and three of their young were spotted one morning by the local postman. Although four of the tigers were caught, including the two adults, one three-year old was shot after being loose for forty-five minutes. The female was observed by Mrs Taylor as it stalked her goat. It then sauntered off towards Littlebourne. One year previous at Redhill, Surey, two strange animals were sighted at a nature reserve and were later captured and identified as coatimundi, badger-sized animals relatives of the racoon. No local zoo had lost such animals and no owner came forward to claim them.

Other theories as to why these cats roam Britain include that they are ghosts of felids from Prehistoric times. This is something that I cannot consider, despite what many peoples opinions are of the supernatural. Black dog folklore is indeed also very potent in the world of the unexplained. Look back through the archives and you'll find many eye-witness reports of phantom hounds roaming Britain. Reports are very much scarce nowadays but were once in abundance. Such manifestations were often described as being Labrador-size, jet-black in colour with shining eyes. These apparitions were often considered as omens of death and misfortune, and were often sighted in dark lanes or on stormy nights. Travellers would often encounter something slinking in the fog that they would call a hellhound. Other popular names for these regional spectres are Black Shuck, Padfoot and Striker, although I have every reason to believe that early reports of these misty canids could well have been black leopards. The behaviour of such forms is more characteristic of a large cat than a roaming dog, unless of course we are dealing with some kind of unknown presence beyond our field of knowledge. However, a majority of 'black dog' encounters describe muscular beasts leaving claw marks on doors, leaping over high fences, panting heavily, prowling near streams, and even leaping from trees to attack. This is not the behaviour of your average stray dog so why should a ghostly dog, if you believe such a thing, act as such?

One theory is that the felids sighted across Britain are of survivors from the Pleistocene era. This possibility seems very remote also and most zoologists tend to agree, because many researchers feel that the explanation as to why so many cats roam the United Kingdom is rather simple, and those solutions have been discussed in this article. Although the lynx may have hung in there, a number of experts tend to agree that the main influx of cats penetrated these shores with the Roman invasion and the numerous other options already spoken of. The only extremely strange fact to emerge out of all this, is as to why no plain spotted leopards are sighted today, or even seem to exist in older reports. Of course, reports dating back several centuries pertaining to large, exotic species in the wilds of Britain, are pretty scarce, but if we look at eye-witness reports from the last century, you'll notice a catalogue pretty much bereft of 'normal' spotted leopard sightings. As already mentioned, reports of larger cats such as the cheetah, tiger and lion may exist in handfuls, but mainly in reference to cats that escaped from a menagerie and then were quickly recaptured or killed. Yet, when we consider how many leopards were imported by the Romans, and also for Royal menageries, it's quite baffling to note the lack of normal leopards sighted among the thousands of eye-witness reports that have been filed over several centuries, even if the normal leopards were quite often housed by exhibitions and private collectors. I'm quite sure that black leopards would have been imported, despite no exact mention of 'black leopards' or 'melanistic cats' in the records, although those vague 'panther' descriptions may well have included, at the time, both the puma and the black leopard. Rather strangely, the absence of sightings concerning normal leopards almost suggests that the whole situation has come about due to black leopards only, being released during the 1970s and '80s, as they were very much a fashion accessory and ego extension novelty pet, but this in turn does not explain all the reports pre-1970.

The African golden cat (*Felis auratus*) is one other cat that could at times be mistaken for the puma and also the black leopard as melanism is recorded in the species. This cat is twice the size of the domestic cat, and could explain a minority of reports of small puma and black leopard in Britain, although those maintained in captivity in Britain would have been few and far between. The natural coat colouration can vary from darkish grey to fawn to reddish-brown, hence its name, as this is a beautiful felid. The backs of the ears are black and the tail has a black dorsal thin stripe, as well as dark bands, which has been mentioned in a few 'puma' reports. Above the eyes of the animal there are pale patches, and the throat, chin and lower area of the cheeks are white. These are powerful animals with longish legs, that feed off small prey in Central Africa. They'll often hunt at dawn and dusk alone, like most cats, but in the United Kingdom reports of such animals would pretty much be non-existent unless such an animal was seen by someone knowledgeable of their cat species.

The Asiatic golden cat (*Felis temmincki*) are said to hunt in pairs in their native habitat which spans from Nepal to Southern China and Sumatra. These animals also have a coat that can vary in colour from golden to reddish-brown, and melanism is not uncommon. On the face, white lines bordered with black run across the cheeks and from the corners of the eyes up to the crown. This cat is larger than the African relation. Again, I'm not sure that such a cat would have been imported and maintained here.

Other cats that may well roam the United Kingdom, but in smaller numbers are the serval (*Felis serval*), which are spotted cats with large ears, and inhabit Africa. These animals have excellent hearing ability and eat small prey such as frogs, rats, fish and birds. The ocelot (*Felis pardalis*) roams parts of the U.S.A., and has a coat of chain-like rosettes. It can reach up to three-feet in length, is nocturnal, solitary and extremely elusive. During the '60s and '70s these cats were extensively hunted for their coat and fortunately this species is now protected but it is still under threat due to loss of habitat. Such an animal would thrive in Britain if numbers were in abundance, and many zoo parks do hold these cats.

Finally, the bobcat (*Felis rufus*) is often confused with the lynx, and inhabits parts of the U.S. and Canada. It is a medium-sized cat with a ruff-like facial border. The coat appearance varies. Spots can be prominent but this density varies, and the general coat colouration is tawny.

All of the cats listed here have been sighted, not as commonly as the leopard and puma however, across the British Isles. Sightings of other cats such as clouded leopards, snow leopards, the jaguar, cheetah, and a few of the more rarer species such as the fishing cat, do exist, but there remains an inconsistency to support their existence in our wilds. Sightings of such cats would suggest that a zoo or private collection is missing a cat, rather than the slim chance that a small number exist in the countryside undetected. And, whilst the newspapers continue to dramatise reports of exotic cats in our midst, and constantly speak of 'Fen Tigers' and 'Nottingham Lions', I believe we should concentrate on the reality, which is far less stranger than the fictions reveals. Although the past remains a little cloudy as to when these cats really began to establish themselves, one thing is sure, that when the license fees were introduced in the '70s and '80s, a lot of people let an enormous amount of cats go into the wilds of Britain.

On 20th October of 2006 *IC South London* reported:

'Panther prowled into my lounge', a bizarre alleged close encounter involving 64-year old Astro physicist Brian Shear from Nunhead Lane, Nunhead, who claimed that during the early hours of the Thursday a big, black cat strolled into his living room and settled on his sofa.

The startled witness told the press: *"It had green eyes and was between four to five-feet long, nose to tail. This was no pussycat. It didn't miaow, it growled. I'd been sitting in the armchair when it walked in. I didn't try to get too close to it because I was concerned it might bite me. I just sat there and talked to it like you would a normal pussycat. I said 'Hello puss, where've you been then?' and it just growled. It seemed quite content and I didn't feel threatened. I don't think it would have harmed me. It seemed familiar with humans."*

The witness, a diabetic, had woken during the early hours after feeling unwell and had opened his front door to let some air into the room. After being in the company of the animal for more than an hour, the creature eventually strolled out the door and headed towards Dulwich.

Could this have been the same animal that had attacked Anthony Holder in Sydenham? Was the animal almost tame, used to humans and a recent escapee from a private collection? Was the creature that allegedly attacked Mr Holder merely an escaped pet seeking human company? In such cases it's difficult to determine what really happened, if such events took place at all. Eyewitness reports are vital to cat research but can be just as frail. Whilst the witnesses remained adamant as to their experiences, such encounters are certainly few and far between, although in Surrey during 1848 an extremely obscure report emerged concerning a man who was found dead by an unnamed roadside. The body of the victim was covered in claw marks, mainly across his face and chest. Rumours at the time circulated that a wild animal had escaped from a Croydon circus although nothing was proven.

On the 23rd October 2006 Peter Dunphy was driving on the M25 on the Kent/Surrey border, between the Godstone and Clackett Lane services when he and his passenger observed a black animal running along the hard shoulder. The creature was cat-like, and described by Mr Dunphy as being three-feet in length and six-feet high but this measurement is altogether too odd. Maybe if the cat was six-feet in length and three-feet in height we could then picture a large black leopard. The animal disappeared into a nearby wood.

At the time of writing (2007), as Spring slips into warmer mornings, the so-called Bexley 'beast' is very much active. On January 5th a woman named Kerry saw a big, black cat in local fields, and within the first four days of 2007 I'd received more than twenty reports of varying exotic species across Kent and the outskirts of the capital after several appeals in the press over the festive season.

During April '07 *News Shopper* stated *'Beast of Bexley could be out there'*, and conducted its own public poll. According to the research, 31.7% of readers believed that some kind of 'beast' was on the loose whilst the same percentage were not sure if the animal existed, whilst 36.6% dismissed the sightings. However, this investigation came just three days after Joanne Parfitt, 30, and Tina Rutherford, 27, spotted a black cat in Erith, which was reported online on the 3rd April.

The neighbours saw the animal in undergrowth near a quarry and went to investigate. They claimed it had a thick tail and appeared to be eating something.

A similar creature was also on the prowl in the Biggin Hill area a few weeks after. An anonymous woman saw scratch marks some twenty-feet up a tree and also heard the deep growl of an animal.
The deputy head warden of Longleat Safari Park commented, *"From what she has described it is possible it is a black panther."*

At 10:20 pm on April 21st, Betty Morris saw a strange black cat stalking the streets of Northumberland Heath, particularly Becton Place. Mrs Morris, 60, was about to close her bedroom window when she saw the animal sauntering along the road. She described it saying, *"...I don't think it was a panther, it was smaller, more like a black leopard and had a huge long tail"*.

Despite her confusion she noted that the creature hid from three men who were strolling on the other side of the road, and then sprayed some bushes with its urine before heading in the direction of some disused garages. The woman then called the local police who probably filed the report.

Whilst sceptics still argue the existence of lynx, puma and leopard in our dark woodlands, and handfuls of researchers fume at their own frustrations at not being able to observe their elusive quarry, it is time to take a step back and realise that the only mysterious ingredient of the British 'big cat' situation is the fact that it's become a mystery at all. The lack of a serious government and authoritive investigation, continued sensationalism from the media, the hardened sceptical attitude, and the camouflaged anorak brigade spewing out mythical statistics at cuddly toy conventions could unfortunately demote the situation into a territory now occupied by the mundane 'Nessie' enigma and UFOs. Let's just hope the large felids roaming the UK outlive them all.

For updates on London 'big cat' sightings visit: www.beastsoflondon.blogspot.com

Sources:

BOOKS

Graham J. McKewan – Mystery Animals Of Britain & Ireland (Hale) 1986
Di Francis – Cat Country (David & Charles) 1983
Jonathan Downes – The Smaller Mystery Carnivores Of The Westcountry (1996) CFZ Press
Karl P. N. Shuker – Mystery Cats Of The World (Hale) 1989

INTERNET
This Is Local London – News Shopper – IC South London – BBC News – Londonist – Beastsoflondon.blogspot.com

NEWSPAPERS/MAGAZINES
Dartford Messenger – Gravesend Messenger – Fortean Times – Surrey Advertiser – Daily Mail – Animals & Men Twickenham Times – Kent Messenger - Daily Mirror

OTHER
London Zoo, Marcus Matthews, The Endangered Species Handbook, & Jemma for continued support and encouragement X

ZOOLOGICAL CURIOSITIES FROM HARDWICKE`S SCIENCE GOSSIP PART 1 1865-1867

Richard Muirhead

Hardwicke`s Science Gossip: A Monthly Medium of Interchange and Gossip for Students and Lovers of Nature* (1865-1893) along with magazines such as *The Gentleman`s Magazine and *The Field* contained the heyday of Victorian curiosity about scientific observation and zoology. Indeed *The Field* is still running today and along with The Countryman provides the occasional gem of Fortean zoology. According to Sheets-Pyenson:

Efforts to diffuse useful knowledge on the part of dedicated social reformers, enterprising publishers, and vigorous voluntary associations created new forms of popular literature in the urban centres of Paris and London during the middle decades of the nineteenth century. [1]

This series of extracts is more concerned with Fortean phenomena than sociological approaches. Charles Fort himself quoted from *Science Gossip* on eleven separate occasions. For example, in *The Book of The Damned*, Fort refers to "a stone that was reported to have fallen at Little Lever, England." [2] This is to be found in the Science Gossip for 1887 p.70. In the summer of 1869 there appeared insects in Britain which had been unknown up until that year. See *Science Gossip* for 1870 p.141 and Fort`s Lo! [3]

The Oxfordshire Crocodile (see p. 104)

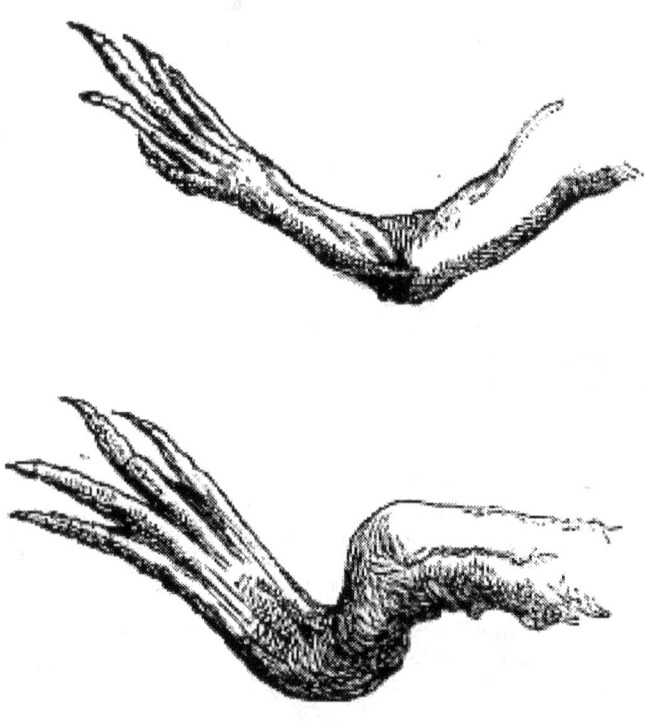

Fig. 2.

Science Gossip was published on the first day of every month, starting with January 1st 1865.

January 1st 1865: A question was posed, " Does the spider eat its own web?"

February 1st 1865: A very long reply was published in the magazine to this question. The respondent only referred to at the end of the piece as `T.K.` commented how in his/her back garden were copious amounts of the garden spiders *(Epeira diadema)* webs. `T.K` cut the edges of one of the webs and proceeded to watch the spider roll up the remains of the web into a ball the size of a pea. The spider then proceeded to digest what remained of the web. `T.K.` wrote that he/she could have dissected the spider to make sure that the web was in the stomach, but this would make no difference to his belief that the web had indeed been eaten.

April 1st 1865: A small communication here from `R.S.B.` which I reproduce here in full: `Your correspondent T.K .has so ably answered the query respecting the garden spider eating its own web, that I can with pleasure bear testimony to the truth, having been a witness to the same nearly forty years back.`

May 1st 1865. An article titled *Viper swallowing its Young*, by A Norfolk Clergyman. This subject was a seemingly endless matter of debate in the Victorian scientific press. The story here is that on an occasion when the clergyman was a boy he and some friends came across a female adder and her young in a clearing in some woodland measuring 20 feet square. Having witnessed the adult snake receive between 6 and 8 young snakes down her throat the lads beat the snake to death. Then

"....one of the party with his knife opened the body, and out came again the little ones, all of which we killed."

May 1st 1865 A white badger. " Lately a very fine specimen of the badger tribe was captured by the gamekeeper on Lord Digby`s estate at Mintern. The animal was of a pure white colour, a very rare species, and weighed over 27lb.

June 1st 1865. There is an intriguing but brief report of a giant spider, measuring, with legs extended, 5 feet long and weighing 6 pounds lowering itself to an " ancient web" in a large church in Lisbon at Easter. It was claimed by an anonymous contributor to Science Gossip to have been a hoax, or was it? More on giant spiders can be seen in Dr. Karl Shuker`s `Extraordinary Animals Revisited` (2007).

June 1st 1865. A contributor named only as `R` wrote to the magazine swearing that he had an account of a viper following a woman who was "suckling a child" home from a wood right up to her cottage door and even attempting to " spring to an upper window." (There is folklore relating to snakes following murderers to exact vengeance upon them.)`R` asks :

"Can it be true that a viper has a predilection for baby's food? Also, is the fat of a roasted viper a certain cure for its bite? Of this I am strongly assured."

July 1st 1865. Another contribution to the debate on vipers/adders swallowing their young: In this intriguing account `H.C.S.` states, that upon coming across a viper in a wood:

" As soon as the viper caught sight of me, it began to hiss, and I distinctly saw several young ones, about three or four inches long, run up to the parent, and vanish down its throat; and from the way in which the parent kept its mouth open, and the young ones glided into it, I should say they were accustomed to that sort of thing."

July 1st 1865. H.A.A. wrote in to comment upon a sighting of a white sparrow which flew back and forth through the smoke of a factory chimney. (Could it have been trying to get rid of ants?)

August 1st 1865. W.E.Williams,Jun.,M.D. contributed his/her observation on white sparrows in smoke, thus:

" I recollect last summer seeing two white sparrows flying through the smoke from the tall chimney of a cloth-mill at Road- a short distance from the scene of the noted Road murder."

August 1st 1865. *Science Gossip* published a communication from a Mr C Gould to the *Proceedings of the Zoological Society* about a group of rabbits he found (but did not say where) playing with a group of young foxes.

September 1st 1865 A white hedgehog. An anonymous correspondent based in Stamford Hill, now in Greater London, reported that " On the 15th of last month (July) , when walking to London, seeing a number of working men examining some curious looking animal......I asked to be allowed to see it also, and to my surprise found it to be a white hedgehog, with red eyes, the same as a white rabbit or white mouse; it was gentle, and apparently very tame, and full grown, and in good health.

September 1st 1865 A white earwig. According to `R.F.M.` a white earwig was found by her/him in a patch of gooseberries. It had black eyes.

" I have preserved it in spirits; thinking it very rare. I thought I should like to know whether it is so or not, and whether any of the readers of Gossip have met with anything of the kind."

The editor of Science Gossip replied that " ...they are occasionally met with."

September 1st 1865: H. Vokes noticed an interesting series of events between two spiders. He destroyed the web of one spider and it retreated to hide under a flower pot for two and a half hours after which time Vokes returned:

…..I found the spider as active as a spider could be, in building a new web, the old one, which at my last visit was still hanging, had now vanished. Had the spider eaten it?- "that`s the rub."

Then a second spider appeared and there was a confrontation between the first spider (spider 1) and the second spider (spider 2). During the course of this confrontation spider 1 collected a large amount of web which

" she began stowing…..away in her own body, forcing it in with her two front claws, and in a few moments not a vestige was left."

Darren Naish confirmed to me in a conversation in the late summer of 2007 that spiders do indeed eat their own webs.

October 1st 1865. Another communication about white earwigs. This time William W. H. wrote:

" I may also mention I have often found colonies of minute young earwigs under garden saucers, and they have generally been *quite white*."

Jan 1st 1866 An anonymous communication concerning a `Crested Blackbird`. Unfortunately as far as I am aware there was no follow up to the following piece of information:

" A specimen of a crested blackbird was exhibited at the last congress of the British Association, which it is supposed may eventually prove to be a distinct species."

June 1st 1866 The interesting subject of adders in Britain displaying unusual colour formations is mentioned by W.R. Tate of Grove Place Denmark Hill. He/She captured "a very beautiful adder" with, in the words of the zoologist Bell, " ground colour almost perfectly white, with all the markings jet black." This adder was deposited "in the Reptile House of the Zoological Gardens, Regent`s Park. Also, Tate adds,

" I was lucky enough to catch him on Wisley Heath,Surrey, on Monday May 7th [1866]

For further information on anomalously coloured adders, see Richard Muirhead`s article `Some Strange Snake Stories` *Centre for Fortean Zoology Yearbook 1998*. Thomas Bell published his `History of British Serpents` in 1839, illustrating a black adder.

November 1st 1866. A tantalising piece of information was published in *Science Gossip* of this date. Titled `Probable Hybrid`, it ran:

" The other day, a gamekeeper brought me an animal which he called " a curious rabbit," and which he had trapped the night before. It is rather larger than a full-grown *wild* rabbit and is of a pale *yellowish-brown* colour, with head rather large, and ears small in comparison. I suppose it to be a hybrid of a rabbit and a hare. Is such a thing of common occurrence? G.B.C.

January 1st 1867 There appeared a long article in this issue on the occurrence in Over-Norton, Oxfordshire, of a young crocodile. This was reported in *The Gentleman's Magazine* of August 1866 under the title " Notes on a young Crocodile found in Farmyard at Over-Norton,Oxforshire" and Charles Fort's Lo! [4] and much more recent publications connected with the unknown. Apparently a Mr Philips was walking in his farmyard at Over-Norton in either 1856 or 1857 when he saw the creature freshly killed, lying in the gutter. Philips expressed regret to the farm labourers who had killed the crocodile and they told him that they could find another at the place where the wood was cut near the `Minny Pool` not far from the village of Salford where they were to be found frequently in the water and often running up trees. The British Museum could offer no definite conclusion as to the origin of the young crocodile . The final conclusion was that the crocodile, illustrated on page 100 with drawings from page 7 of *Science Gossip* from January 1st 1867, had escaped from a travelling menagerie. The crocodile measured, from the tip of its nose to the end of its tail, 12 to 13 inches, from the tip of its nose to the crown of its head about 2 inches and the front legs about 1 and a half inches and the hind ones about 2 inches long.

March 1st 1867. The following story will be familiar to students of Fortean Zoology. At Falkingham,Lincs in October 1866 a hen was deprived of her eggs which she had been sitting on in readiness for their hatching. They were taken away from her, but in their place a female cat gave birth to her kittens, which were literally "taken under the wing" of the hen, who proceeded to nurse them. " During this singular attachment she would always make room for the old cat to suckle them."

Correspondence with the author of this article can be made to:
Richard Muirhead
112 High Street,
Macclesfield,
Cheshire,
U.K.
SK11 7QQ

or richmuirhead@tiscali.co.uk

ON THE TRACK OF ORANG-PENDEK?
by Nick Molloy

Some long-term readers of *Animals &Men* may remember my last venture. Graduation had brought new found funds enabling my mission Loch Ness armed only with a tent, a backpack full of cereal bars and cheese and my much maligned girlfriend in tow. Then year was 1997.

The year was now 2004 and much had changed. The world of work had brought moderate wealth only for me to lose it all in a failed business venture. A year in exile in Dubai returning to my dreaded occupation in recruitment allowed me to recoup significant lost funds. However, a combination of the appalling nature of the Dubai lifestyle and working for wankers ensured that I was never going to take the offer of an extension to my year's contract. This meant that under UK tax law I had to remain out of the UK for a few months longer. Some suggested I stay in the sandpit, they simply didn't understand. Others suggested I migrate to a beach in Thailand for a few months. I couldn't think of anything more dull. Instead, a few months traveling to interesting places ensued..........

My much maligned and still ever-present girlfriend is still in this tale. Her chosen destination was New Zealand where we spent nearly six weeks exploring both islands in our hired 4x4. It is hard to believe that a country with such a varied fauna can be housed in such a small space. If you get the chance to visit this wonderful country, take it. You won't regret it. My chosen destination was Nepal. I have always wanted to explore the high Himalaya and find out a little more about the land of the Yeti. However, the yeti is not our quarry here. Sandwiched between our two chosen destinations was a month in Indonesia. The first two weeks in Indonesia were spent on the tourist Island of Bali. To be honest this was the dullest part of our four month excursion. Even staying in a five star beach hotel, I quickly became bored with the lack of action. I went scuba diving, climbed a volcano and even visited a Komodo Dragon and got up close and personal with a 13ft Python. Ultimately however, I ended up spending money each day hiring jet skis ands seeing how long I could get airborne such was my level of boredom. Beach holidays just don't do it for me.

The next stage of our journey took us to Sumatra on the track of Orang-Pendek. I had long suspected that the elusive Sumatran ape-man was cryptozoology's best bet. By that I mean of all the alleged cryptids out there, many of which I believe have plausible explanations and are little more than myth and legend, the Orang-Pendek or small man, may just turn out to be that real flesh and blood creature that one day walks out of the forest and becomes mainstream. By belief was further fuelled by a couple of expeditions that went in search of Orang-Pendek (including our own Richard Freeman) and came back with promising reports. Due to me being in Dubai, I unfortunately missed the talk by Adam _____ at Uncon 03 on his trip to the West Sumatran jungles. Although the reports I received back from my spies posed more question than answers regarding their findings.

I therefore decided that the only way top find out more was to travel there myself. I should point out at this point that the only planned event in the entire four months was our airfares. We had no contacts or no accommodation at any of our destinations. We just turned up and went from there.

I knew from previous readings and a brief conversation with Richard Freeman roughly what we would have to do on arrival. I knew of Debbie Martyr and her Orang Pendek crusade so I considered her a good contact to find.

We flew into Pedang, the main city in Western Sumatra. It was just getting dark as we arrived. Pedang had the most oppressive weather of any place I have ever encountered. Being right on the equator it was warm, but the humidity combined with the smog made breathing difficult. It made the Dubai summer with its 50 degree heat seem moderate. Pedang was not a welcoming sight and rather than hang around we decided to head straight for Sungeh Penuh in the hope of finding Debbie Martyr and a guide. Richard had mentioned an uncomfortable eight hour bus ride. We decided to hire a taxi and were hopeful of making it there for midnight. Pedang is not a place inhabited by Western tourists and English speakers are few and far between. We found a taxi driver willing to make the trip for approximately 60GBP (he spoke virtually no English) and we set off at about 6pm.

What ensued was the trip of nightmares. We travelled for about three hours up twisty, windy mountainous roads surrounded by thick jungle. Suddenly we hit a convoy of traffic all pulled into the side of the narrow road. People conversed with our driver, chattering away excitedly in the native tongue. We knew not what was going on, but ever so often we would progress a few more yards. In the end I got out of the car and went to investigate. After walking for some time it became apparent that the rains had washed away part of the road. There seemed little we could do except to sit and wait, which seemed to be the general consensus. Many passing Sumatrans all stopped by our cab to stare at the unusual white faces they contained within. They pointed and gesticulated at others to come and see the spectacle. It was a slightly unnerving experience, doubled by the presence of my girlfriend. The hustle and bustle created by excited Sumatrans continued unabated outside our car. After approximately another three-four hours a local hot shot turned up in a toned-down monster truck and using sign language offered to pull us through the patch of the road that now resembled a mud slide for a small fee. We duly accepted and a few scary minutes later we were on our way again. We reached Sungeh Penuh some 12 hours after our journey began. I also had to persuade the driver to pull over and take a nap or risk him driving us off the road ! He took us to a hotel, where no-one spoke any English, nor was there any apparent interest in our custom so we went onwards to another. It was now 6am and we had been travelling for nearly 24 hours. I was not enjoying Sumatra at this stage.

We went onward to another hotel where by a stroke of luck a guest who was checking out spoke both English and the local tongue. He was an American teacher present with a couple of colleagues. I told him of our mission to find Orang-Pendek. Despite having been in the area for a couple of years he had no knowledge or had even heard of such a creature. He told me where I could find the Kerinci-Seblat National Park Information Centre and informed the hotelier of our wishes for a room. We crashed and slept for about four hours. The temperature was unpleasant and the room even worse, but this was not to deter me from the mission.

We rose and set off to find the Information Centre. As we were walking to it's location we quickly became aware of the incessant staring and pointing at the white man and his woman. It was very clear that tourists don't really permeate this far into Sumatra. To our dismay the information centre was shut because it was Sunday. I mentioned the name Debbie Martyr to a few people outside but only got nods which clearly meant nothing. A man who spoke a little English found us outside the building. He had been told by others of the lost looking whites roaming the town. I explained as best I could about the quest for Orang-Pendek and our need for a guide. He beckoned us to come and meet his friend. He led us over a busy street and down a side alley into a small shop and bade us sit down whilst he fetch his friend.

He returned within a few minutes with a man called Afnir who spoke good English. He explained that he worked for Kerinci-Seblat national park and could help us find a guide. He was very welcoming and commandeered a couple of scooters to take us to his office and then onwards to a local eatery. We discussed our requirements and I mentioned Sahar who had been recommended by Richard as his guide. Afnir informed us that Sahar was not available, but there were others who were his equal who were. We paid the agreed fee and were taken to meet Yan (our guide with a good grasp of English) and Pen (our porter who doubled as a guide but spoke no English). Yan was unusually tall for an Indonesian, (circa 6ft) and Pen was unusually short (lucky if he was 5ft). Indeed Afnir joked that there was no need to ego into the Jungle as Pen was Orang-Pendek, or short-man.

Afnir drove us to a homestay for that night which had home comforts such as a shower (albeit cold). The homestay was a short drive from out start point and was directly opposite Mount Kerinci, an active volcano and at 12,500 feet, the highest mountain in Sumatra. We had discussed with Afnir our plans. It was agreed that firstly we would hike up to the lake Gunung Tujuh (seven peaks) as this represented and area of many supposed Orang-Pendek sightings. We would spend three nights exploring the area surrounding Gunung Tujuh. We would then return and make an assault on Mt Kerinci, because it had been reported that an Orang-Pendek had stolen the rice from a German tourist from outside his tent. Further details on this incident were lacking. Also, a climb to 12,500 seemed like good training for Nepal where we would be going considerably higher.

The next day we rose early and caught a bus to a local town. Yan then marshalled us on to the back of a couple of scooters to reach our start point marked by a small hut. We registered our name in a book inside the hut and began our trek into the jungle. For what was the next three and half hours we climbed steadily along a well worn 'path' carved out of the jungle. The weather was hot and sticky and both Donna and I were carrying packs as well as Yan and Pen. Donna found the going tough, but I certainly wouldn't label it as overly difficult, either in terms of terrain or gradient. I was carrying an altimeter and from our start point at just over 3000ft, we climbed to a maximum height of 6540ft where we encountered a small decrepit shelter. The air was certainly getting cooler and crisper as we rose and by the time we reached out high point the temperature had dropped noticeably.

Donna did find the going tough and after a couple of hours, Jan insisted that the indefatigable midget, Pen, take her pack. Being a stubborn Scottish lass she refused, but did at least let him take a couple of the heavier items. Yan was constantly Jolly and upbeat and made a noticeable attempt to help here through the difficult bits. Richard also told me that he really struggled with this section of the climb.

I guess I should mention at this point that I am an ex-International track sprinter and still keep myself in shape for my new profession (male stripper). Although my speciality wasn't exactly long endurance treks I still found the climb only moderately difficult. I emphasize this point, not to illustrate my own fitness or 'big up' my own ego, but simply to compare the terrain with what was to come. To those not used to training, the climb may be difficult, but to those that have a sport as a passion it probably wouldn't be overly taxing. It is perhaps best compared to ascending Ben Nevis on slippy mud. Indeed, the path resembled a giant staircase made from tree roots and branches. When I plodded this route again a little over a week later, it took comfortably under two hours.

From the decrepit shelter we descended just over 300ft to the shores of Gunung Tujuh at 6200ft. The skies turned grey and it began to rain. The lake stretched as far as the eye could see and it was surrounded on every shore by the sight of thick, impenetrable jungle. There was the roar of a waterfall nearby. Yan explained that we would canoe along the lake to a nearby hut where we would spend the night, he would just have to attract the attention of the fisherman. In the distance a body got into a canoe and began to make it's way towards us.

As it neared the shore Yan said that it was Sahar (Richard's guide). They chatted in their native tongue and Yan explained that Sahar was acting as a Porter for Lee, the senior guide, both of whom had other clients at the hut, so we would have to pitch a tent. I tried to interrogate Sahar about previous expeditions, but he appeared uninterested. We climbed into the 'canoe' - for which, read: hollowed out log - and began our precarious trip across the lake. I had to sit very carefully so as not to rock the boat, so to speak, so that we wouldn't take a swim. We canoed for about 20 minutes before reaching the huts, only for me to find that I was cramping up given the position I had been sitting. We were introduced to Lee the senior guide and a couple from Brighton! I enquired why they were in Orang-Pendek country. They replied that they were just traveling throughout SE Asia and had not even heard of Orang-Pendek until Lee had told them about it. Indeed, they were amazed that I knew so much about it.

My lasting memory of that day is one of serene tranquility. As the rain beat down on the lake and the mist rose off the surface of the calm water, I remember staring into the distance and the surrounding jungle, wondering what secrets they might contain. After a dinner of noodles, we gathered around a fire in one of the huts. I began to talk to Lee the senior guide. He had lived in the region all his life and was now 47 years old. He no longer carried the loads he used to, but was clearly well respected by the other guides. I asked him whether he believed in Orang-Pendek. He claimed that he had seen it when he was a boy, but never again. I asked him whether he could have mistaken his Orang-Pendek for a Sun Bear. He very calmly, matter-of-factly said 'I don't think so, I know bear' When in business I made my money from reading people and situations. I found Lee to be quite genuine and found no reason to disbelieve what he was saying. He certainly seemed to believe he saw Orang-Pendek. I also asked him if he had ever seen a Tiger. He replied that in all his years in the jungle all he had ever seen were its tracks. He didn't seem to be trying to impress us with his tales of his exploits. Lee was to be the most impressive supporter of Orang-Pendek that I encountered.

On the walk up to Gunung Tujuh, I had been interrogating Yan on his beliefs of Orang Pendek. As he was acting as a guide for an Orang-Pendek seeker he drew a balanced picture stating some believe in Orang-Pendek whilst many believe it is a myth. It was obvious at that point that he was a non-believer. Afnir, had also trod a similar path the day before when I interrogated him. I attempted to interrogate Sahar on the same subject but he seemed disinterested and I got the impression that he only responded with what he thought you wanted to hear.

The next morning we rose to brilliant sunshine and I felt a warm glow in the morning light gazing outwards on our panoramic vista. Despite the fact that we were clearly in a remote part of the world I didn't truly feel it. Not too far down the jungle path I knew there existed a third world civilization. I had also felt similar feelings of serenity at

Scottish and Irish lakes. I somehow expected it to be different.

After breakfast we again boarded our hollowed out logs and begin our journey to the very end of the lake. For the next hour I scanned the shores of the flat calm lake for any signs of critters as we neared our destination. Eventually we pulled into a small clearing no more than 100 yards from the far end of the lake. It showed signs of previous human habitation. There was a primitive shelter with four wooden posts and a piece of canvas covering its top. However, the canvass was torn and one of the posts had fallen. The clearing was completely overgrown and despite the rapid rate of growth out there in the jungle Yan estimated that it was over a year since the last humans had come this far. We immediately pitched camp and set up our tents.

After something to eat we began our trek into the surrounding jungle. There were remnants of a previous explorer; a vague route could sometimes be distinguished, clearly slashed back with a machete sometime in the recent past. However, this was completely different to what we had encountered before. Yan knew the area; he had ploughed this very route before although not for some time. I suspected it was a route that all the guides use to bring the stupid westerners who seek adventure in the jungle.

Whereas our route up to Gunung Tujuh had been fairly straightforward, now we were really in the bush. In some places the vegetation was so dense you were surrounded on all sides and it felt like you were ensconced in prison cell of green no larger than a phone booth. In one area with a little more breathing space Donna took a photograph of me. I am crouching down slightly as an Orang-Pendek might do (see photo 1). I am no more than 20 feet from the camera, probably only 15 feet, and as you can see I have virtually disappeared. I began to think that a large creature could walk right past us and our chances of seeing it would be remote.

We came to a stream. Yan like a goat danced over some fallen logs some eight feet or so above the brook. The athlete in me marvelled at the dainty, graceful way in which he negotiated the obstacle. He had clearly done it before. We on the other hand were not so skilled or graceful. I began the traverse with a little trepidation. Balance was never a strong point of mine. I reached the half way point across the divide when suddenly the wood gave way under my feet. It felt like I was falling through a trap door in a medieval castle. It happened so suddenly and there was nothing to grab to stabilize myself. I instinctively reached out to grab something. All this served to do was turn me sideways. I crashed through the wood and hammered my left side on a log as I fell. I then fell back and hit my head of another log as I hit the damp floor. I regained my feet somewhat sheepishly and fought my way out of the myriad of rotting wood. My head stung and I checked for blood, but there was only a protruding lump. I felt my rib, but I had taken harder shots than that in the gym so thought nothing of it. It became transparent within a couple of days that I had fractured the rib.

We pressed on, I was feeling a little sheepish after my fall and didn't say anything for a few minutes. I was struck by how quiet the jungle was and I figured we must have been disturbing the wildlife. I spoke to Yan and asked him to continue without me. He seemed a little puzzled, so I explained. I wanted to simply stand next to a tree, quietly, not making a sound so that our heavy footprints would not disturb the inhabitants. I asked him whether he could find me again if they all left me just standing there, because I had no idea where our camp or even where the lake had disappeared to. The jungle was all-encompassing and had completely disoriented me. Yan assured me he would be able to find me again and said he would return within the hour. Donna gave me the camera in case anything interesting might wonder by.

Within five minutes of being left alone the jungle quite literally became ALIVE. From an eerie silence it erupted into a crescendo of noise. Monkeys began chattering, birds started calling and the surrounding undergrowth began cracking and creaking, presumably as birds flew in and out. I stood there for a full half hour watching, listening for anything out of the ordinary when the noise again abated and familiar human chattering emerged from the undergrowth. Yan was early, presumably a little concerned at my strange request to be left alone in the middle of the Sumatran jungle.

We began our trek back to camp. My altimeter was not enjoying the Sumatran weather. Hot and wet conditions were causing it steam up when exposed to the air. I guess it prefers the cool mountain air.

That night we lay in the entrance to our tents staring up at the stars, warming our hands by the fire and chatting incessantly about our different cultures. All was well, but things were about to turn nasty.

We rose to a gray ashen sky, the kind I have always admired. The jungle had acquired a menacing hat to its lofty tops and had a moody air. We photographed a couple of purple crabs in the lake and took many photos of the mist rolling off the lake in a scene reminiscent from 'The Fog'. We packed up our tents and broke camp. Our plan was to hack through the jungle back to where we camped on the first night, searching for evidence of our quarry on route.

Donna had left her pack at camp 1 on the advice of Yan. We were due to be away only one night after all. I still had my pack with my now fractured rib, the pain for which was starting to become more apparent. This manifested itself in shortness of breath and sneezing/coughing became very painful. It had hit home by this stage that my rib wasn't just bruised. Not to be deterred by such trivia we set off into the dense undergrowth.

I distinctly remember that morning because it was the first time we had been exposed to such difficult terrain whilst I was carrying a pack. We started to ascend again, but this was nothing like the trek up to Gunung Tujuh. The going underfoot was slow and difficult. We were constantly having to traverse difficult sections, clamber over fallen trees and in some cases underneath fallen trees on our hands and knees. Underfoot was slippy and muddy making balance difficult. Yan and Pen both seemed to possess a goat like quality meaning that they fell very rarely. I seemed to fall every other step. Furthermore, every time you lose balance, your body reacts naturally in an explosive manner to try and regain its balance. Your heart rate soars to feed the increased blood and you naturally start to fatigue at an increased rate. Yan and Pen's bodies were functioning much more efficiently in this terrain with the humid conditions. They didn't fall and they didn't seem to fatigue. I for the first time was finding the going difficult. I am sure my damaged rib didn't help, but I won't use it as an excuse. The terrain and the going were difficult.

The rains started again at midday as they always did, you could set your watch by them. I looked at my watch, which was steaming up again. The height read 6800ft and Yan informed me that we would now be descending back towards the lake. I took off the watch and put in the pack in an attempt to protect it from the humid conditions.

We descended for about an hour. My rib became quite uncomfortable and I began asking Yan how far the camp was, his response was always not far. After two hours of descent I was beginning to fear the worst and I asked Yan if he was lost. He denied it, but failed to convince me. We were crossing a stream and I lost my footing yet again sending me careering into the water. I rose looking at my scraped hand and noticed what looked suspiciously like a leech on my right knuckle. I called out to Yan and he confirmed my suspicions.

Yan had assured me that we would encounter no leeches around Gunung Tujuh as it was too high and too cold. I was wearing what I considered to be leech proof clothing none-the-less. I looked like a track sprinter in winter; I had skintight leggings and a skintight top below my waterproofs. Everything was tucked in so I had very little skin exposed. My hands were one of these areas.

I had read that it was unwise to pull leeches off and was better to either burn them off or apply a little salt. I asked Yan if he had any cigarettes left. He didn't. How far to camp I asked, ten minutes he replied. I figured I could let this sucker have its fill for another ten minutes. However, I had started something. Everybody started noticing that they had leeches on exposed parts of their anatomy and unlike me, they were all wearing loose fitting clothing.

A further half an hour passed and the bloodsucker had actually moved along my hand. I decided it would have to come off. I pulled at its bulbous little body only to lose my grip and have it slap my hand like a stretched elastic band. This time I gripped it a little harder and it came loose only to then stick to my finger. I had to scrape it off on a tree. Donna was starting to get a little freaked about the bloodsuckers that were attacking her. I again challenged Yan on our bearings, he assured me he knew where he was, but |I was beginning to fear the worst.

After another half an hour, I removed the rains finally stopped and I removed the watch from the pack, long overdue. To my horror it read 4400ft, nearly two thousand feet **below**

Gunung Tujuh. I halted Yan and explained that he was lost. He again tried to re-assure me, but I explained that we were now over 500m (he wasn't comfortable with feet) below the lake and that continued descent was only taking us away from our goal. The full horror of our predicament hadn't yet hit me. I still had this instinctive mis-placed confidence in our guide. Maybe the watch had finally blown a fuse in the humidity and was giving a wrong reading. After all it was never 100% accurate. However, as we trudged on, the realization slowly began to hit me. The most the watch had been out before had been 200ft. I know that regular altimeters start to fail above about 13,000ft, but as yet it hadn't been that high. It couldn't possibly be that far out. Besides it was still registering minor changes in altitude every two minutes, as it should do. Also, we had been steadily dropping for a long time and even given the difficulty of the terrain we had clearly dropped more than 600ft in that time. I don't know whether with a combination of fatigue and rib pain I had just failed to register the obvious, but it hadn't registered.

After another half an hour or so, Yan finally conceded were lost. We had also run out of water and we had no food as we left all of our rations at camp 1. The realization of our predicament finally hit me full on like a Marvin Hagler right hand. I felt like Roy Scneider in that famous scene from Jaws when he sees the shark for the first time. I guess shock best described my state. I turned to Donna and mumbled that I was scared. We were lost in the world's second largest expanse of rainforest. We had no food, had just run out of water. We had no GPS, no-one knew where we were, we could see no more than a few yards in any direction and our best chance of escape was our guide who was just as lost as we were.

We were up the proverbial shit creek without a paddle. Also, it began to occur to me that as we were now in a survival situation, it might benefit Yan and Pen if they were to conveniently lose us. They were far quicker over the terrain and would have a better chance off escape without us. My general impression of Yan did not suggest this, but I had known him barely four days and the rules of the game had just changed. I did not mention this thought to Donna.

My much maligned-girlfriend seemed in an equally shocked state, but the realization of our predicament had clearly not hit here. Her shock was arising from the creepy crawlies sticking to her skin. There was about an hour of daylight left. I suggested to Yan that we pitch camp. He eventually agreed. We found a flattish area and tried our best to clear it of undergrowth and pitched our two tents in the centre of it. Yan suggested that Pen go and search for some water as we were all parched. I wasn't so sure. Even though we were in a rainforest and water couldn't be far I thought he may lose our position. Anyway, off he went.

Donna, now near hysterical with the thought of leeches began to strip off. She was covered in them. Yan pulled over 20 of the evil little fuckers from her (he had the knack). They have a strong anti-coagulant to prevent your wounds from clotting so they can suck you dry. Donna looked like an accident victim. My leech-proof lycra had done the trick. I kept catching them crawling up my leg having failed to gain entry around my ankles and boots. They reminded me of a cross between miniature giant sandworms from *Dune* and John Carpenter's *The Thing*. They resembled the sandworms in look but they way they whipped the heads around in search of bare flesh reminded me of The Thing.

Donna retreated to the sanctity of the tent. Pen had been gone for over 10 minutes and Yan began shouting him. There was no response. This continued for another 20 minutes until Pen finally showed up. He had got lost, but was fortunate enough to wonder into earshot and follow Yan's calls back to our clearing.

We settled down to sleep not long after dark, tired from the day's exploits and anxious over our situation. Donna seemed to have a misplaced dislike of the leeches. However, it seemed to keep her mind of our predicament.

We awoke at first light (6am) and began to break camp. Pen and Yan had been discussing strategy as had Donna and I. Yan was off the view that we needed to retrace our steps upwards in an effort to find the lake. I that this was self-defeating. I favoured finding a river and following it down and out of the jungle. I based this on a very basic knowledge of the local geography. A mountain range runs virtually the entire length of Western Sumatra. If we could follow a river for long enough we would eventually encounter the coast road and could flag down help. We may also encounter some sort of settlement next to a river. If we had water we could survive a good couple of weeks and this should give us sufficient time to reach the coast road. Logically, despite our predicament, I felt that this strategy offered us a 90% chance of survival. I felt that a search for the lake could last a lifetime and cost us our lives.

I asked Yan if there was any fruit we could pick from the jungle. He said there was bananas but they weren't in season. We were going to go hungry then.

In support of his plan Yan tells me that Pen believes that he saw the abandoned animal tower that we encountered on the way up to Gunung Tujuh. I tell him this is not possible as the animal tower was at 4800ft and we are now at 3900ft. I was unsure how to take this story. Was Pen trying to reassure himself regarding our situation, did they not believe the altimeter?

Yan suggested I give him some time to see if he could retrace his steps from yesterday and if not we would follow my plan. I agreed and we set off. For about an hour we rose and was continually turned back by the dense foliage. It soon became apparent this wasn't going to work, so Yan conceded defeat and we began to follow a small stream downwards.

The jungle was so thick and steep sided that we could not actually walk along the river, so we began wading through it. It was never more than knee deep to start with, but the going was slow. We soon began to encounter small waterfalls that we had to negotiate carefully. At least we had water. This continued for about the next four hours. Our stream became a river, widening and deepening. Soon we were wading through a gorge with steep sided foliage on either side of us. We came to an enormous waterfall (about 40ft in height). We weren't going to jump off and there was no other apparent way down. We were going to have to climb again and hope that there was an alternative way down.

Yan suggested that he and Pen go on a reconnaissance on the left side of the gorge to try and find a suitable route. I agreed, but suggested they leave their packs for us to guard (I was still conscious they may attempt a runner). They agreed, leaving us to rest on a rock whilst they went route finding. It was midday and the rains began again.

Yan and Pen returned about half an hour later stating that there was no route through. There was no choice for it but to climb on the right side of the gorge above the waterfall and hope for the best. We began our precarious ascent with caution. We were climbing on 45 degree slopes with minimal grip and a big drop over a cliff if things went wrong. I was at the rear making slow progress. Suddenly I lost my footing on the left, the right foot scrabbled for grip where there was none. My bottom half began to slide, I shot out my left arm and managed to grab a small tree to arrest my fall. My body continued to fall towards the cliff, but luckily the tree held and my arm held my weight, my right foot was dangling over the abyss. I pulled myself back up and continued the ascent. I had torn the rotator cuff in my left arm, but adrenaline is a wonderful thing. I never felt it for a couple of days and the pain in my rib had also miraculously abated.

We reached a ridge and began down the other side. The roar of a large river soon came into earshot. However, the descent became steeper and more frightening. I was still at the rear. Donna was ahead of me and Yan and Pen were descending in their goat-like fashion at the front. Suddenly Donna fell and began to slide down the slope at speed. Within a second she was gone from my view. I began to run and call out to Yan. I soon lost my footing and began sliding down the slope, face down, at considerable speed, clawing at the mud with my fingers to try and arrest my fall. After what I estimate was 3-4 seconds of rapid freefall I cam to a halt. Donna had come to a halt at the same place and had managed to stick a leg out bringing me sliding to a halt on the slope. We were perched on a steep sided slope with nothing but foliage continuing into the distance below us. We managed to regain our footing and continue our descent. It wasn't long before we emerged safely, if a little shaken by the banks of a river. It had doubled in size since we left it at the waterfall. Indeed, we couldn't even be certain it was the same river. It also became apparent that if we hadn't arrested our last fall we would have fallen over a small cliff into this river. The fall would only have been in the region of around 12 feet. It certainly wouldn't have been life threatening, but it could have caused a nasty injury, such as a broken leg and that could have signalled the end of the game.

We stopped for a rest by the riverside. We had a small amount of salt and sugar left

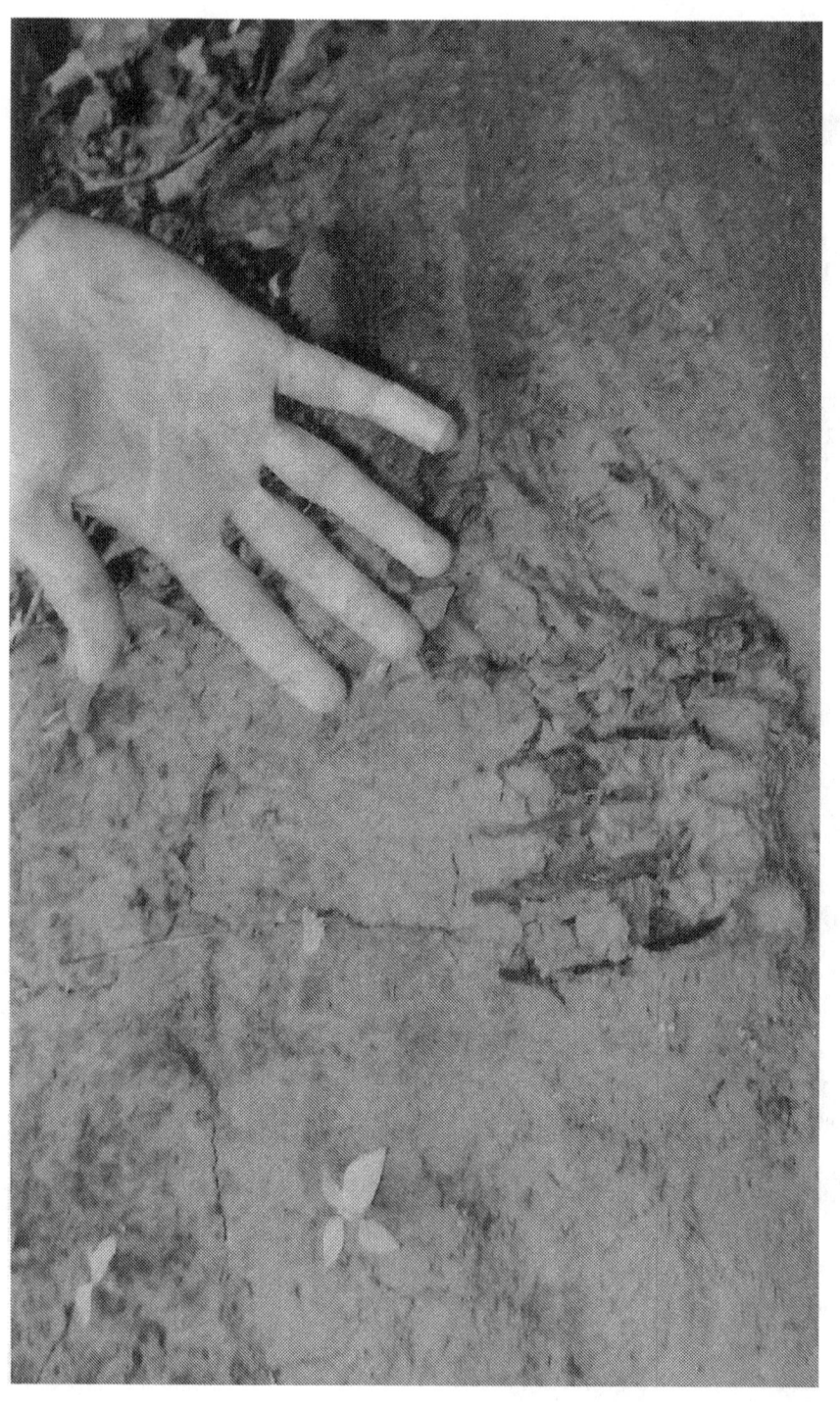

Left - a picture of my hand next to what I think is a fresh tiger print. The more imaginative may say it is a Cigau. I say Tiger.

I was 10 yards from the camera and took this before getting lost. It was intended to illustrate how dense the foliage is and how easily a man sized object could be missed even only a few yards away.

over from our rations (but no actual food). We began mixing this into our water in an effort to increase out hydration and help prevent cramps. It was becoming apparent to me that the constant struggle to maintain an upright position in this difficult terrain combined with a lack of food was noticeably draining us. It occurred to me that in subsequent days we would probably be unable to maintain the current pace due to increasing weakness.

Your body instinctively knows it is in trouble in this sort of a situation. Your metabolism slows and begins to burn its fuels stores slower than normal. I figured we were going to need it.

We set off again wading through this much larger river. At points we were wading waist deep and the river also turned into rapids in parts. The going became very slow. The jungle showed no sign of thinning out. There was one point where we had to clamber over some rocks and cling to a rock race to escape the rapids below. I tackled this second behind Pen. Donna later said that she feared I would be washed away as I clung to the rock race.

However, we escaped yet another precarious section and emerged to a calmer more serene stretch of river. I distinctly remember at about 3pm that the sun came out and the rains stopped and we stopped for a rest on a sandbank as the river curled round to the right. I remember thinking at the time how nice the situation would be if we weren't lost without food ! A small cliff of clay and mud looked across at us from the far side of the river where a deep pool had formed. As I lay there in the warm sun, I almost fancied a swim !

We set off again and for a while progress improved, but the rapids were to return periodically. By 6pm the light was again beginning to fade and I said to Yan that we would again have to think about pitching camp. He said we could keep going and that Pen thought he could see a clearing in the distance. The panic was beginning to show in his voice. I said that whatever was there would be there tomorrow, we had to rest and recuperate for another day. Yan climbed about eight feet up the near vertical riverbank, using his machete to dig into the mud to give him purchase. He said there was a small area up there and we could pitch the tents their. He helped the rest of us up and we made camp as best we could. We got out of our sodden clothes and made ourselves as comfortable as possible.

That night my body really began to stiffen up, I bean to fear for our progress. I couldn't move for my rib and breathing was difficult. I just reasoned it would improve when I started moving again. My hands were torn to shreds and stung like hell. Sleep was a blessed relief.

We again arose at dawn. It was raining, something it hadn't done that early before. We were down to 2200ft, so we were making progress. When Yan emerged from his tent, he burst into tears. He said he was scared he might never see his daughter again. When the guide says that you fear the worst ! I tried to reassure him. He kept apologizing for getting us lost. I said we needed him more than ever now, as he was our best chance of getting out alive. He recomposed himself and we broke camp and set off.

The jungle began to thin out a little. We were able to make our way through the undergrowth in parts, although we had to keep criss-crossing the river to enable ourselves to do this. We just kept hitting impassable walls of foliage. After two hours of this we encountered fresh tiger prints in a muddy bank. We also shortly afterwards came across what may have been at one time a human path. Things were looking up. We had a difficult decision. Follow the path or stick to the river. We decided to follow the path and hope for the best. After 10 minutes it came to a dead end and we had to retrace our steps. This happened twice more. We continued along the river and soon came across and old abandoned shelter. We had dropped to 1700ft and everybody had increased optimism. Our route along the river widened out and soon we were following what Yan interpreted to be an old logging trail.

We were continuing along this path when we heard the sound of a chainsaw. It's whirring blades made a sound sweeter than music. We hurried to the source of the noise and emerged above a small but fast flowing stream. On the other side, two wiry looking Indonesians were at work on a felled tree. Yan waded through the stream and began chattering excitedly to the two men. He translated that the men were part of a camp of illegal loggers based about two kilometers from our current location. Yan had assimilated the directions from the men. I handed over some $$$ for the helpful advice and after Yan assured me he wouldn't get lost again we set off through the jungle.

True to his word, Yan soon found a very rough-looking track that was wide enough for a logging truck and we set off uphill. Donna and Yan set off with renewed enthusiasm ascended the much easier gradient. I on the other hand, perhaps feeling less threatened, suffered an adrenaline drop that saw me trudge slowly up the hill, nursing my wounds. Pen seemed to have the same ailment.

Due to my trudgery, I missed a rather exotically coloured (blue) snake swallowing a rat. By the time I had reached the site, it had slithered away to safety in the surrounding jungle.

We walked for around half an hour before, true to their word, we came across what can be loosely termed a camp. Rough wooden huts had been erected around a clearing that was clearly man made. A few beasts of burden stood around lazily chewing on the grass. A couple of motorbikes were the only intrusion of modern life. Someone came out and was soon followed by a deluge of others. Yan again began chattering in Indonesian. His conversation continued for a couple of minutes before he turned to me and ushered me to show my hands to the strangers. He explained that he had told them we had come through the jungle from Gunung Tujuh and had got lost, but, they didn't believe him ! They said that it would not be possible for man to trek on foot through such harsh terrain! They did have a point; only fools or the desperate would attempt it.

When we showed them our hands, cut to ribbons as they were, their suspicion abated a little. Yan told me that there was a logging trail which if we followed for another 60km or so would take us out of the jungle. Sometimes trucks would sometimes reach this far into the jungle, but as we were in the transition between the wet and dry seasons, the trail was too muddy. Only bikes were reaching this far into the interior. Our new hosts tolds us we were 80km from Gunung Tujuh. I have no way of checking the authenticity of the claim, which I will come back to later.

Yan suggested we remain at the camp for the next day in the hope that a truck may yet reach for the morning. This seemed a forlorn hope to say the least, but as we were tired, wet and hungry it didn't seem a bad plan. As the sun came out we tried to dry some of our clothing. Yan discovered yet another leech, but with time now on our hands took a sadistic revenge, cutting it in half with his machete.

Our new hosts gave us tea and noodles. News of our appearance from the jungle had spread rapidly and the whole camp had now appeared to stand and stare as we hungrily devoured our noodles. It was more than a little unnerving. Personally, I couldn't care less, but the presence of my girlfriend with a couple of dozen strange, staring men put me a little on edge. Perhaps sensing my uneasiness, Yan tried to alay my concerns. He told me that they had never seen white people before. We might as well have been Martians who had just landed in our spaceships.

We were led to a room in one of the wooden huts and lay down. My rib was now feeling like it had taken a few digs from Sugar Ray Leonard and was resisting sudden movements. Our room had no light except from a wooden shutter that served as a window. What little light was brought in from outside was soon diminished by a gathering crowd at the window who just stood and stared at us as we lay there.

The afternoon became evening and soon our room began to fill with our hosts. A couple of them were from one of the towns and came into the jungle to work. They spoke a little English and were keen to practice, as all Sumatrans appeared to be. Candles lit the room until well after dark and we chatted despite our weariness. I again brought up the subject of Orang-Pendek. The general consensus was that it was myth not reality, although this view was majority not unanimous.

Yan asked me if I was ready to sleep, to which I replied in the affirmative. He politely informed our hosts and within seconds the room was clear. I wondered outside for a call of nature. As I wondered off into the bush, I was again followed for more group staring.

Despite the hard wooden floor, I slept like a baby. I remember waking because a large rodent was running across one of the beams (Yan lit a torch and said something like shuu to it). The next time I was awoken was from the noise of the shutters being flung open at first light so that more people could group stare as we awoke.

Obviously there was never going to be any rescue truck, so we set off on our long walk after breakfast. We paid our hosts and thanked them for their hospitality. Our elevation was such that we now felt like we were on the equator, the heat was really quite nasty and we had to stop for regular water breaks. We were now following a winding, but never ending muddy trail. Jungle existed on either side of us, but it was flatter, not so mountainous and less dense, partly due to logging.

We continued for hours over dips and rises, passed by the occasional motorbike going to or from the camp. At one point we encountered an abandoned truck where it had got stuck trying to ford a stream.

We encountered several side forks to the main track, but continued along the main route. Eventually, it became harder underfoot the lower we got and our hopes of a lift out of the jungle rose. The indomitable Pen was finally looking mortal on this more normal terrain. Removed of his jungle terrain advantage, he started to look more human and the deep cut on one of his feet couldn't have been helping. He spent so much of his time in the jungle wondering in bare feet and it was only when the leeches began to attack that he put on his socks and sandals !

An empty truck came from behind us and Yan flagged it down. We clambered in to the back and clung on for dear life as we began a bumpy ride along the track. Also, the driving left a lot to be desired and didn't instill confidence ! He only took us a couple of miles to another fork in the road and told Yan that if we walked the other way we would arrive at another pick up point shortly.

We crossed a bridge over a now much wider river and came upon a couple of wooden huts, one of which acted as a 'restaurant' presumably for passing loggers. The restaurant owner told Yan that he had heard of Orang-Pendek sightings locally. Other than that information was sketchy, but he was a believer.

We waited for a couple of hours and sure enough at about 4pm caught another truck. This was loaded with planks of wood, but was destined to leave the jungle. However, we had a 30km drive ahead and it was destined to be an hour of pain! I almost began to wish I was back in the jungle. Still, the dirt tracks finally gave way to a partly concreted road and the backdrop of jungle and increasingly wide river valleys was a stunning vista given the now setting sun.

We passed through small villages and paddy fields. Our driver passed through checkpoints where someone emerged from a nearby hut and took payment. Yan looked at me, shrugged his shoulders and said just one word – corruption.

We continued onwards and were finally dropped off at a small town, where we had a short wait before catching a mini-bus that passed through daily, we had been lucky with our timing. Another couple of hours and a short walk later we arrived back at our homestay.

There was some concern for our whereabouts, as we had not returned on time. Afnir was apparently mobilizing guides to launch a search party. They had gathered something was amiss. However, my confidence in them was not high and my suspicions were supported (more later)

It was clear that we were not now going to climb Mount Kerinci, we were fast running out of days. Yan said that the next day he would go and fetch our bags from Gunung Tujuh as they were of course still at the hut. I insisted on accompanying him and the next day we set off together. Unburdened and with the easy terrain, we reached Gunung Tujuh within a couple of hours. At the top, just above the lake we encountered a party of school children who had been camping out at the lakeside. They informed Yan of two German tourists who were lost in the jungle with their guides. He didn't bother correcting them.

We returned back down the mountain. Yan informed me that he was feeling very guilty about getting us lost and by way of repayment he would like Donna and I to come and stay with him and his family for the remainder of our stay in Sumatra (now only a couple of days).

We agreed and were treated well by Yan and his wife. I thought this was a really nice gesture and was something he didn't need to do. We spent our last couple of days at Yan's home on the out-skirts of Sungeh Penuh. Before we went to stay Yan kept telling us that they were poor people and not to expect anything special. I think he was a little embarrassed and somehow expected that we lived like royalty and would be disappointed by his home. It was rustic by western standards, but still comfortable and homely. Of course all water has to be boiled and there are no sit-down toilets. Other than that, there wasn't much to miss.

Staying with Yan, gave me an opportunity to visit the Kerinci Seblat national park visitors centre on a day that it was open and begin to pour over what maps they had of the area in attempt to pinpoint our position. It soon became clear we were off the map. Where we were was marked green and that was it. The only other details were the rivers. This I was assured was an accurate depiction of the terrain taken by satellite. However, nobody was able to tell me where (at least accurately) where we had emerged from the jungle. The road to the logger's camp did not exist on the maps, thanks to their illegality.

Therefore with only the main road to go on and a sense of the distance covered I was able to pinpoint only our general location within the jungle. To my horror, it was obvious that we were on the wrong side of the mountain range to reach the coast road. Although the rivers were dropping, they were doing so in the wrong direction from our point of view. If we had continued to follow the river in the jungle, we would not have reached the coast for a few hundred miles, as we would have had to follow them through to the Eastern seaboard! You may recall earlier that I had given us a 90% chance of survival when the bad news struck. With this new-found data, at the time I would have dropped that figure to about 60-70%. All of a sudden I felt like we had been lucky to escape as opposed to having done the right thing in the wrong situation. Maybe we would have followed the river and reached a settlement, but maybe we wouldn't have. It seems just as likely that we would have simply walked into more uncharted, less inhabitable thick jungle and we might never have come out.

The question arose in my mind as to whether we entered the so called lost valley 'where no human has ever been'. It certainly seems possible, even likely. However, my supposition would be that there are several lost valleys in the jungles of Western Sumatra where no human has been. Certainly when I met up with a very relieved Afnir upon our return to the homestay, he said to me that we couldn't have paid for the journey we had just had. None of his guides, even Lee, had journeyed that far into the jungle and he wouldn't be recommending it either. I am certainly not of the belief that the 'lost valley' harbours unknown creatures and that all we need to do is make an excursion into it and we will find a colony of Orang-Pendeks. I think this is the notion of overzealous cyptozoologists and does little for the cause. All of which brings me onto the subject of the elusive Debbie Martyr.

Whilst at the centre I also took time to study some of the plaster casts footprints of the creature alleged to be Orang-Pendek. I do not know enough about primate prints to critically evaluate what I saw or even to critically assess the possibility of them being fakes. There was little information surrounding the prints (where they were found or by whom), but on what I saw could not be called irrefutable evidence.

There were also a couple of articles penned by Debbie Martyr lying around or pinned to the walls. When I read these, I am afraid to report that my heckles rose and some of my suspicions were confirmed. I had found the stories surrounding Debbie as somewhat incredulous. A journalist who comes out to Sumatra and has a life changing experience and doesn't go home. That's all well and good and she may be a lovely person. Where I am struggling is her relationship with Orang-Pendek. I don't understand how experienced guides and local people who have spent all their lives out in the jungle, yet have failed to ever catch sight of an Orang-Pendek (or a tiger) Yet, a western journalist arrives, spends a limited time in the jungle, yet sees it three times. Given the ability of the local people on the terrain (they are certainly nimbler and less clumsy) a find her fortune all the more astonishing. Furthermore, I was also told the story of another western researcher who spent considerable time in the area placing trip wire cameras in the jungle in the hope of catching Orang-Pendek on film. He trekked into the jungle everyday to check his cameras. He never saw anything, and his films never revealed anything unusual. Debbie Martyr, must possess the scent of Orang-Pendek since she seems to attract them so much! Either that or she has misidentified what she has seen or perhaps in the tradition of journalism has stretched the truth a little in the pursuit of a good story.

She had certainly done that in the articles of hers that I read. It stated clearly in one of them that if it was suggested to a local that Orang-Pendek didn't exist they would think less of you (I'm paraphrasing) and see you as a madman. This is an outright and blatant lie. I was asking everybody who would speak to me (either directly or through an interpreter) whether they believed in Orang-Pendek. I was struggling to find people who did! Furthermore, these were not city people, these people lived in the small villages that bordered the jungle. They could not be described as jungle natives, but their beliefs could not be dismissed as the naivety of city folk who didn't understand the countryside. Many of them made their livelihoods by living off the jungle and its environment.

You may be beginning to gather that I think Orang-Pendek is nothing but a myth. However, this is not the case. I simply abhor this type of sensationalist journalism, particularly in a field such as ours. It does nothing except give the naysayers petrol for their bonfire and allow them to label us all as daydreaming legend-chasers. There is something there that is worthy of investigation. Yet, if the investigation proves that Orang-Pendek is nothing more than a myth, then the investigation does not become any less worthy. It merely allows us to put that matter to bed and move onto the next investigation. The facts should speak for themselves and our job as cyrptozoologists should be to report them as such. If people want to create fiction, they should do so under the correct heading.

So what has our trip to the jungles of Sumatra revealed? In short not an awful lot. For me, the three main theories all remain possibilities. That is firstly, that Orang-Pendek, is an as yet unidentified primate roaming Western Sumatra. Secondly, that Orang-Pendek is merely the sighting of misidentified Sun Bears and thirdly that Orang-Pendek is actually Orang-Utan. This latter theory is born from the fact that Orang-Utan used to inhabit the West of Sumatra within the last couple of hundred years, although it is now confined to the North of the Island. Some theorists postulate that sighting of Orang-Utan by natives a couple of hundred years ago, have been handed down by generation after generation, and somewhere along the way Orang-Pendek has been born. It is certainly easy to see some of the descriptive similarities (short in height, immensely strong, red hair).

Interestingly of note, all of the people I met who were believers or claimed to have seen Orang-Pendek were from the older generation. Those who claimed a sighting also interestingly claimed it from their youth, NOT currently or recently. It was the youngsters who were dismissive of Orang-Pendek. This raises the further notion that maybe Orang-Pendeks were more abundant in Western Sumatra 20-30 years ago and have come under population pressure, possibly due to the increased logging activity. They could have died out or possibly moved into remoter parts of the jungle. This may account for the notable reduction in belief and reported sightings.

There is certainly still enough jungle for large unknown cryptids to remain hidden from prying human eyes. Sumatra contains the world's second largest rainforest (only the Amazon is larger) and parts of it are completely unexplored and very remote. For me, the possibility of Orang Pendek remains very much alive.

On a final note, whilst we stayed with Yan and his family after escaping from the jungle, he insisted on showing us a video that he kept referring to as 'rescue on Kerinci mountain'. Yan is a member of Kerinci mountain rescue, a team of local guides who are meant to come to the aid of those who are unfortunate to get lost on the mountain (no helicopters here though). He told us that some Javanese got lost on the mountain and this video was made of the rescue.

Yan had a battered old VCR and TV on which to show us this masterpiece. On the night in question, many of Yan's friends came from far and wide to see yet again the rescue on Kerinci Mountain. The tape whirred into life to reveal some actors being filmed on camcorder marching towards Mt Kerinci. It wasn't long before the footage switched to 'live'. There were shots of Kerinci rescue headquarters with the leader of the rescue team excitedly chattering and drawing the plan of attack on a whiteboard to his thronged mass of rescuers.

The footage cut to a scene in the jungle. There were rescuers scattered everywhere, lying lazily smoking cigarettes and chatting. I asked Yan had they found them yet. He said not yet, this was camp 1 at xxxxxft. The scene was repeated shortly afterwards at xxxft – camp 2. There was no urgency about the rescuers. Indeed, they looked as if they were enjoying a day out in the forest.

The scene cut away again, this time the surroundings were different. Harsh rocks jutted out from outcrops. There were no trees only bracken. The height must have exceeded 10,000ft because they were above the treeline. Yan said that the rescue was imminent. I was on the edge of my seat, when suddenly the camera cut to a body, a very dead body. Then, it cut to another body and yet another. Rescue on Kerinci mountain was a body retrieval !

I knew we had no hope of being rescued when we were lost in the jungle. The terrain was too vast, too mountainous and largely impenetrable. Without a GPS, any rescuer would have been looking for a needle in a haystack and was just as likely to end up lost themselves. Rescue on Kerinci Mountain simply took my view to another level.

However, all this has not put me off. Fully aware of all the risks and having had a truly frightening experience, I would go back for more. The threat of the jungle and the fragility of our guides has not put me off the search. I went to Sumatra believing that Orang-Pendek was a probability. I am afraid that I have come away from Sumatra with a view that at best can be described as agnostic verging on the skeptical. This view has been born out of poor reporting by the cryptozoologists. If there was genuinely a creature there and the evidence to support its existence there would be no need to hype its proposed existence. The clear fact that its very being has been hyped now leads me to question the foundation of the evidence on which we have been led to believe. That is, the main reason to create hype is to sell something which frequently is bogus, fraudulent or sub-standard. The fact that Orang-Pendek has been hyped is not a good advertisement for Cyptozoology or its proponents. I now suspect that if many of the original sources for claimed Orang-Pendek sighting were traced to the source, what we have read and have been led to believe may not be all it was painted to be. This is not to decry Orang-Pendek as fiction or non existent. Yet, to conclusively summarize, I am now far less of a believer than I was previously and I feel like it's my own side that has scored an own goal.

MADNESS MONSTERS AND MORAR
Lisa Dowley

In the spring of 2005 a small three person CFZ team (Richard Freeman, Dave Curtis and myself) undertook a small field and fact finding expedition which was kindly funded by Dave Curtis, to the remote highlands of the west coast of Scotland to gain a basic perspective of the lay-of-the-land and to test out some ideas we had regarding flotation devices, in respect of an alleged creature that resides in Loch Morar, one of the many Lochs in the Highlands which is less well known than the renowned Loch Ness, but whose folklore and legends have endured none the less.

The actual name Morar has its origins in 'Mordhobhar' which is Gaelic for 'big water', and big water it is indeed, as this is the largest body of fresh water in Europe, it is in fact a remnant from the last Ice Age and is a very deep sided glacial lake which has never been known to freeze, it is over eleven miles in length and in some places over a mile in width. The deepest part of the Loch has been recorded at over one thousand feet deep, and the clear waters themselves are rather productive playing host to a variety of life.

The Loch itself does not flow directly into the sea, as it falls short by a few hundred yards at its western end. The actual link to the sea is made by one of the shortest rivers in the country: the River Morar. At around half a mile long, this short rivers life begins with dramatic gradually-dropping forty foot falls which flow out into the silvery sanded, blue watered bay.

This is combined with steep sided hills and a mountainous peninsula, which appear from a distance clothed in velvets of green and purple hues which at random seem torn by scattered outcrops of rock add to this the remoteness, and welcoming sound of true silence and you genuinely have a place of outstanding yet strangely eerie beauty.
With such a picturesque backdrop it is not surprising that it has been utilised by film and TV companies in such productions as *Highlander* and *Monarch of the Glenn*.

This idyllic yet remote location is populated by some two hundred Morar residents, which have a very strong sense of community, which is reflected in the fact that they are all party to many local secrets and legends, most of which they are reluctant to discuss for fear of ridicule. The clear freshwater depths of the loch harbor many secrets and stories that can instill fear and disbelief and are of equal to any of the fearful stories of the Highlands.

Stories such as the 'Colann gun Cheann, this is reported to be a headless woman specter which was thought to kill passers-by that unwarily traveled by night near the railway station which was formally known as the 'Smooth Mile'. If that was not enough to unnerve you while wandering through the village of an evening, how about 'An Cuth Glas Mheobail', the 'Grey Dog of Meoble', this spectral dog was greatly feared. There are many versions of this particular story; however they are all based around the killing of a bitch and her young puppies by their master. It is reported that this large spectral hound with its luminous blood-red eyes can still be met at night in the woods that lie between Arisaig and Morar.

However the legend that is most familiar and the one that has drawn the CFZ team to this remote part of the Highlands, is the story of Mhorag, (pronounced Vorack in Gaelic). This Mhorog was the spirit of the loch it could appear as a large serpentine creature or present itself as a woman of fair beauty. However caution is needed as seeing the Morag was said to be an omen of death for a specific Highland family.

Morag may not be a creature of fable such as a kelpie or a water bull, as mixed in with such tales of myth and folklore are stories that seemingly stand out due to their more factual and descriptive manner, recollections of a creature with an eel-like head, which on occasion has been seen with and without a humped back, occur semi-regularly and with eerie consistently.

Morag, the monster of Morar, has been sited on numerous occasions over the centuries and was even mentioned by St. Columba, however the beast did not seem to appear or gain much notoriety until the second half of the twentieth century.

This could have been for a variety of reasons; the elders of the Morar community would over time, have died and taken their stories and descriptions of such unknown and feared creatures to their graves and as such who knows what valuable information may have been lost.

Also, the majority of small clusters of what community there was, that lived on the shores of the Loch, have abandoned there Loch-side dwellings most probably in favour of a more appealing, favourable and less harsh a lifestyle in the cities.

And so it seemed that sightings of large unknown creature(s) all but cease until after the Second World War, the Loch then seems to steadily bubble back into life, with the idyllic appeal of the scenery, it is not long before the area is re-discovered and the advent of tourism brings people and boats, back to the Loch and as you would expect reported sightings begin to increase and be documented for posterity.

In July of 1948 a sighting was reported in various newspapers of the day. A Mr. John Gillies, was making a living by taking boat loads of tourist up to the head of the lock and back on daily excursions, he was more than familiar with the lock, having some twenty-four years of experience of this body of water under his belt.

On this particular evening it was around 7pm and the boat with nine passenger tourists onboard, was approximately half way home, off the southern shore 'between Meoble and Eilean Allmha, by all accounts it had been a warm sunny day, and the Loch was as calm as a millpond.
One of the passengers first noticed some odd movement in the water about a quarter of a mile astern. Having heard the commotion John Gillies came out of the engine room and looked through his binoculars. The general consensus of what was being observed by witnesses was around twenty feet long, with five humps, which were described by one witness as reminiscent of 'five salmon backs'. It was stated that these humps were not undulating in a coil like manner of a possible sea serpent but by all accounts appeared ridged, neither a head nor a tail had been seen, however the creature was said to move slowly which caused a slight wash and remained in sight for over five minutes.
This is the only report that states there were more than three humps visible, Mr Gillies could have mistaken the other two, possibly being caused by the creatures wake, but in light of Mr. Gillies experience as a boatman and his familiarity with the Loch it seems an unlikely mistake.

Ten years later there was another group sighting, albeit from a family.

In September 1958, Dr and Mrs Cooper along with their two children were on holiday at Morar, had an encounter.

Dr Cooper, a keen amateur artist was sketching while the children were playing on the shoreline. While sketching he became aware of what he took to be a submerged log drifting very slowly towards the west, he described it as a 'flattened S', course drifting'. This he observed for around half an hour, making sketches while doing so, at one stage stating that it was around 'some fifty yards' from the shoreline in relation to the Cooper family. When he looked up from sketching the scene before him, the object had disappeared; however his daughter said that it had vanished in a 'swirling mass of water'.

In recalling a description of what he saw, he thought the object to be grey-black in colour, approximately twelve feet in length, and its shape was a 'very flat ellipse', but thought the shape changed gradually during its slow movement. The actual texture of what he saw, to him seemed rather rough such as a tree trunk, but definitely not shiny.

Over the years many sightings were reported occurring between the months of July through to late September, this could be expected as it being high in the summer season and there would be more individuals out and about on the Loch, although there have been sightings documented during the colder winter months.

One of the most dramatic accounts occurred on August 16th 1969. Duncan Mc Donnell and William Simpson were returning from a trip up the Loch. It was around nine in the evening but still light. Mc Donnell was at the wheel and the boat was doing seven knots. Here is the series of events in Mc Donnell's own words...

"I heard a splash or disturbance in the water astern of us. I looked up and saw about twenty yards behind us this creature coming directly after us in our wake. It only took a matter of seconds to catch up with us. It grazed the side of the boat, I am quite certain this was unintentional. When it struck the boat it seemed to come to a halt or at least slow down. I grabbed the oar and was attempting to fend it off, my one fear being that if it got under the boat it might capsize it."

Simpson's recollections of events were as follows...

"As we were sailing down the Loch in my boat we were suddenly disturbed and frightened by a thing that surfaced behind us. We watched it catch us up then bump into the side of the boat; the impact sent a kettle of water I was heating onto the floor. I ran into the cabin to turn the gas off as the water had put the flame out. Then I came out of the cabin to see my mate trying to fend the beast off with an oar, to me he was wasting his time. Then when I saw the oar break I grabbed my rifle and quickly putting a bullet in it fired in the direction of the beast... Then I watched it slowly sink away from the boat and that was the last I saw of it."

Neither of the men seemed to think the bullet had any effect on the monster. They estimated it to be some thirty feet in length, (nine meters). The skin was rough and dirty brown in colour. It had three humps that protruded eighteen inches out of the water. McDonnell thought they might have been undulations rather than humps. McDonnell reported seeing a snake like head a foot across held around eighteen inches out of the water.

With such a dramatic encounter it is no wonder it drew the attention of a small body of scientists and biological researchers, who turned their attention from the well known Loch Ness, to Loch Morar and so in 1970 the first of two scientific based surveys (the other being 1971) were undertaken on the Loch. The two surveys were later condensed and presented in a book which included a concise history of the area, detailed accounts of sightings, the aims, methods and conclusions of the survey, and was aptly entitled '**The Search for Morag**' by E.M. Campbell and D. Solomon.

These two surveys were headed by Elizabeth Montgomery Campbell a former member of the Loch Ness investigation and David Solomon, who holds a degree in Zoology from Exeter University.

Together with other team members they incorporated such techniques such as visual surface observation, electronic aids such as sonar to assist in underwater observation and various biological testing of the environment. It was even documented that some members of the Loch Morar Survey team witnessed a large black hump-shaped lump, and then a large body of disturbance in the water.

They came to the conclusion that there appeared no reason why a large predatory species of animal could not be supported in the Loch, and further went on to suggest various species, which may have given rise to the various sightings, but were reluctant to pin-point the mystery sightings to one particular creature or species, but rather preferred to give a list of possibilities, and so concluded to only ever scientific based survey of the Loch.

Then over thirty years later in April of 2005, the CFZ sent a three person team to this remote outpost of the Highlands to put to the test some 'bait holding flotation' devices that the resident cryptozoologist, Richard Freeman had designed, in line with his theory and belief that Morag is indeed a giant eel.

We arrived in Morar just as dusk was beginning to fall and the madness of the long, arduous seven-hour journey was soon forgotten as we turned into the drive of the bed and breakfast, which was to be our home for the next few days. Waiting to greet us was a beautiful peacock welcoming us with his tail in full display. This we found out to be one of the many pets full of character very much like their owner, and proprietor of Garramore Guest house, Julia Moore. Indeed Garramore proved to be just as mysterious and full of history in its own right as the Loch itself.

It was built in 1840 as a hunting lodge, then during World War Two it was used to train spies that would be parachuted into France, rumors of double agents being found out, shot and buried in the grounds are abound in the area, and indeed in Julia Moore's own words…

"I tend not to dig too deeply around the shrubbery in case of what I might find".

Indeed Julia Moore proved to be a fascinating individual in her own right, her links to the highlands go way back in history as one of her ancestors was none other than Lord Lovett who was a rebel catholic supporter of Bonnie Prince Charlie, who hid out on one of the islands on the Loch (Eilean Ban) smashing all boats in the area thinking he would be safe. He was wrong.

After being discovered he fled but was later captured. The then aged Earl was marched on foot to the Tower of London where he was to be executed a year later. During her youth, Julia lived in London, and had such neighbours as Quentin Crisp and also knew the renowned author Mervin Peaks. She was a hive of information and related to us a sighting that until our visit had not been recorded and was unknown outside of the village.

Apparently two youths from Yorkshire were on a fishing holiday about four years ago. They were on a boat, one keeping watch the other operating the tiller. The lad on watch shouted out there was a 'tree' approaching the boat at an alarming rate. Both saw what looked like a tree trunk racing towards the boat. They feared a collision but at the last moment the 'trunk' arched up and dove down into the depths. The pair made for the bank in record time, packed up their tent and returned home on the same day.

The evenings at Garramore were spent round the open log fire, Richard, David and myself along with Seal, Julia's pet collie dog and Badger her pet cat, for comfort, discussing various theories of what Morag may be.

These theories ranged from the time old plesiosaur concept through to Richard's Freeman's view, which is that it's quite possibly a giant sterile eel.

He summed up his theory as follows…

"These are common eels that swim out to the Sargasso Sea to breed then die. The baby eels follow scent trails back to their ancestral fresh water homes and the cycle begins again. Sometimes however a mutation occurs and the eel is sterile. These seem to stay in fresh water and keep on growing. Known as 'Eunuch eels', no one knows how old they can get or indeed how large one can grow. Just last year in February 2004, two Canadian tourists came across a twenty-five foot eel floating in the shallows of Loch Ness. At first they thought it was dead but when it began to move they beat a hasty retreat".

But just how plausible is this theory of Richards?

Indeed on closer inspection there seem to be many conceivable connections that would suggest that Morag could well be a giant eel. Eels are very illusive and little-understood creatures and where they spawn was not discovered until the twentieth century.

They do start life in the Sargasso Sea, (near Bermuda) and it is thought that they die after spawning; however, due to the tremendous and unattainable depths at which it occurs, on-one has never witnessed this act.

After spawning the eel larvae drift in the warm Gulf Stream via the Atlantic where they reach the shores of Europe. This journey can take several years to make and by which time the larvae will have turned into a green eel known as an Elver.

These Elvers then follow their genetic instincts and head for freshwater, where they spend some time in the river estuary preparing themselves for their transition from salt to fresh water, a short river such as the River Morar would be an ideal place for this transition. Eels can also travel over land as they have the unique ability to absorb oxygen through their skin as well as their gills, it is also thought that they can close their gill flaps to trap a small amount of water inside, thus covering their gills and enabling them to travel further via land, some have on occasion been found in fields a considerable distance from water. So if the gradual waterfalls of the River Morar prove too much for an eel it would be more than possible for it to make the move via land. Again this would give support to various sighting that have been made on the shore and marshy edges of the Loch. It is thought that eels will spend five years plus growing in their fresh-water home as they are rather a slow growing fish, before succumbing to the genetic urge to return to the Sargasso to spawn, but for as yet unknown reasons some eels decide not to make the return journey, preferring to stay in their watery home for in excess of twenty years. One can only imagine what the appearance of say a thirty-foot-plus mature eunuch eel could look like!

There have been many and varied description of what the creature of the Loch actually looks like. Ranging from snake-like to having the appearance of a dogs head! Again the remarkable capabilities of the eel come into play; it can quite literally change the shape of its head. On arrival at their freshwater home the eel will have a small pointed head with a small mouth which is most suited to feeding on small invertebrates such as snails and worms, if there is nothing more substantial for the eel; its head will remain this shape. If the eel elects to settle in a large body of water such as Loch Morar, where there is a large abundance of fish and water fowl etc, it will change its jaw shape to suit its predator instincts, the head becomes broader in shape and the teeth become rather prominent.

As for the varying in difference of colour and patternations recorded in the many sightings of the 'Monster of Morar', one possibility for consideration, is that eels are thermophilic, meaning they prefer warm water, and they are shy of high light intensities. Preferring to Hide in deep mud and weed during the day. In the summer month's eel's move into the muddy shallows to feed as the light fades. A giant eel that has woken from a torpid state and covered in mud, weeds and other forms of Loch vegetation, which may have dried in some parts due to basking in the late evening of the summer sun's warmth could give the appearance from a distance, of something other worldly and indeed monstrous to the human eye.

Over next few days our thoughts were pre-occupied with giant eels and we were more than eager to get out on the Loch, however this sadly was not possible as we could not get in contact with the only local man who hires boats. While becoming acquainted with the Loch and its surrounding landscape, we were able to bare witness to just how quickly and choppy the waters could become.

Julia had warned us about how sudden and violently quickly the change to inclement conditions could be. Just last year a young man had decided to go out on the Loch In such weather and had disappeared. His upturned boat was found adrift. Despite police searches with divers and heat detecting devices the body was never found. It is common knowledge that in cold lakes decay in bodies is slowed by the low temperature; consequently gases do not build up to buoy the body to the surface. The unfortunate man's body may have sunk down to the bottom. On the other hand something else could have happened to it...

While surveying the Lochs shoreline we were rather surprised at the amount of dead sheep washed down from the surrounding mountainous hillsides and ending all but in the Loch itself. These would make a first-rate additional food source for a big predator in the Loch with a keen sense of smell such as a giant eel.

During our observations of the shoreline we were able to pick out an ideal spot to test the flotation devices. We used nylon rope with empty plastic milk bottles as floatation devices. We tied one end to a tree, one of many that had sprung up from a rocky outcrop. We then lashed another length to the bottle connecting the two lengths. At the end of the second length, forty feet or so beneath the float was the bait. This was a mixture of mussels, fish guts, herring, cow liver and Van Den Eynd Predator Plus, this being a fish-attracting chemical. The mixture was placed in Hessian sacks so it could permeate and diffuse through the water with the hope of such a scent attracting a large eel.

We returned to this snaggy and woody outcrop just off the main shoreline the following day hoping that something may have taken a nibble at the smelly, fishy concoction that only an eel could find tempting but sadly nothing had touched the bait, in spite of this the floatation devices had worked a treat, and had remained in tack and were considered a success as regards incorporating them in future investigations of a similar nature. Our general consensus of opinion was that it might have been too early in the year as the creature may still have been in its winter torpid/semi-dormant state.

We hope to return to Loch Morar in the New Year, preferably in the late summer months, as this would coincide with a considerable number of reported sightings. It is hoped that we will have a larger team, which will enable the CFZ to carry out proposed experiments and more in-depth research in a bid to determine weather Loch Morar is indeed inhabited by a Giant Eel?

- CFZ YEARBOOK 2008 -

SOURCES OF INFORMATION

'Richard Freeman'
Cryptozoological Director of the CFZ.

'The Search for Morag'
By Elizabeth Montgomery Campbell and David Solomon

'Julia Moore'
Proprietor of Garramore Guest House

'The National Anguilla Club'

SOME NEW ZEALAND CRYPTIDS
Tony Lucas

Laughing Owl - Hakoke, Whekau

Sceloglaux albifacies albifacies (South Island)
Sceloglaux albifacies rufifacies (North Island)
Once the maniacal laugh-like call of this bird rang through the night forests, then in just 40 years this call was heard no more. However, reports persist.....

The laughing owl was moderately sized – its height being 14 to 15in, and with a wingspan of 10 ft 4 in. It had reddish brown plumage streaked with darker brown, and a white face.

The North and South Island birds were sub-species, which only called while on the wing, and these were mainly heard on dark, drizzly nights, or preceding rain. The South Island birds were larger than the smaller North Island species; males were generally smaller than females.

Abundant until around 1845, within 40 years this charming little bird had disappeared. However, the call of the laughing owl has been heard often since, and there are those that believe it may not be extinct as thought.

This species preferred open country for hunting, and rocky areas for shelter. The rocky areas of the Southern Alps were very much suited to its needs, as were areas of Canterbury and Otago. The birds showed a preference for low rainfall areas of the country – the areas of Nelson and Fiordland were also favoured by these birds, and remains were found on Stewart Island in 1881.

In the North Island, they were said to inhabit the Hakoke Cliffs, as well as the Ureweras, where they lived in holes in the cliffs, in the upper reaches of the ranges.

A clear indication of their diet has been gained from the plentiful fossilised pellets that have been discovered, and it would appear that the laughing owl fed on lizards, insects, and small birds. It was a ground feeder, with sturdy legs, that preferred to run its prey down.

The nests were made of dry grass and were generally constructed on bare ground, and in rocky crevices, where two white eggs were laid.

This bird was known to the Tuhoi People in Te Ureweras, in the North Island. Birds were also said to be found, in pre-European times, in the Albany area near Timaru.

A North Island bird was collected from Mt Egmont in 1856, and Wairarapa in 1868; and around this time birds were also reported from the Porirua area, and Te Karaka. Mr W.W. Smith managed to breed some of these birds in captivity in February of 1882, and several fine specimens, along with eggs, were dispatched to Buller, together with letters describing the breeding behaviour and care.

July 1914 saw the last sighting of a laughing owl; a specimen was found dead at the Blue Cliffs Station in Canterbury. The only physical proof remaining of these birds were 57 type specimens, and 17 eggs, in public collections (Worthy 1997).

It seemed however, that the laughing owl was not totally through. Unconfirmed sightings of laughing owls came from the North Island in 1925, and in 1927 one was supposedly heard at Lake Waikaremoana, when it flew over giving a weird cry, almost maniacal in nature.

In the 1940s, a laughing owl was reported in the Pakahi, near Opotiki (Parkinson), and a sighting was witnessed at Manpouri in 1950.

In the South Island, in February of 1956, eggshell fragments were found at Saddle Hill in Fiordland, but the most recent hope for this species came in 1960, from the Canterbury region, when what appeared to be reasonably fresh eggshell fragments were found.

Various expeditions have been mounted to try and find the laughing owl, but the results have often been inconclusive - there have been possible calls heard, and occasional pellets and egg fragments found, but never any glimpse of this elusive bird.

Why these birds became extinct is somewhat of a mystery.

Their decline over 40 years has puzzled many. It is believed that the invasion of weasels, stoats, and cats may have spelt their doom. Rats were no problem to these species, as they actually provided a new food source for this bird - as evidenced from pellets that have been found. Whatever the reason for their decline, unconfirmed reports still continue to come from areas such as Fiordland, and various areas of New Zealand. But still no photos, or live birds. Perhaps, in the remote areas of Fiordland, the damp night sky still rings with the maniacal laugh of this enigma.

Hopefully perhaps the laughing owl may have not, yet, had the last laugh.

New Zealand Moose

Moose (*Alces alces andersoni*), were imported from Saskatchewan Canada, into the South Island by the New Zealand Acclimatisation Society as a sporting animal, along with red deer, in the early 20th Century. These animals supposedly failed to establish, and yet sightings persisted, and evidence of their continued presence continued.

AAHFAB Alamy Images

The initial introduction occurred in 1900, when four animals from Canada were released in Hokotika. The initial release was supposed to have been fourteen animals, but ten died on the voyage from Canada. Out of these four animals only one was a cow, and was said to wander the streets of a local settlement until 1914, when it was no longer seen.

These animals were presumed to not have survived, and a further release was planned. This occurred on the 6th of April 1910, when 10 ten-month-old calves (six female and four male) were released in Supper Cove in Dusky Sound.

It was believed these animals died out due to the competition from red deer (*Cervus* elaphus). However, a small number must have persisted as reports of physical traces, and sightings continued. These sightings be-

came quite prevalent between 1929 and 1952.
Herrick Creek was one spot where a bull moose was supposedly shot by one Eddie Herrick in 1934.

This was one of a dozen animals shot between 1910 and 1952.

The last one sighted, and shot, in 1952 was presumed to have spelt the end of the establishment of a moose population in New Zealand - in fact it would have been the only population of wild moose in the Southern Hemisphere.

Nothing more was heard of the moose, apart from rumour and speculation, until a possible sighting in 1971 sparked a hunt for a potential surviving population of these enigmatic animals.

More physical proof came to light in 1972 when an antler, which was definitely from a moose, was found. That year, on the insistence of the New Zealand Forest Service, Ken Tustin was charged with finding out if they still existed. Research conducted by him suggested that a small population might have inhabited the Dusky Sound area.

This was based on prints, droppings, antler casts, and signs of grazing. However, no actual sightings of the animals were forthcoming - the thick bush in the area kept them well hidden.

Mr Tustin did not give up the search, and in 1995 a picture was taken of a possible female at Herrick Creek. A single frame from a video clip showed what appeared to be the retreating figure, but, regrettably, the image was not a clear one due to the video being in time-lapse mode, which causes some distortion to the picture. The outline and stance of the animal, however, were quite convincing.

A hair sample found in 2000 was subjected to DNA testing, the result of which confirmed that it was definitely moose hair. Further evidence of their continued existence was found in 2001, when two hunters came across some more hair. This sample came from Shark Cove, on the southern side of Dusky Sound, and was also confirmed as moose hair.

In October 2002, another hair sample was found, opposite Oke Island, by Mr Tustin, which was snagged at waist height on some tree bark. This, too, was subjected to DNA testing, and proved to be of moose origin.

Though no actual sightings have occurred since 1971, there is plenty of physical evidence for the existence of these animals. However, it would seem that these animals appear to be confined to the Dusky Sound area.

Bedding spots and physical evidence suggest that up to 20 animals may live there (*Otago Daily Times* 06.10.05). Although the population is small, these animals seem to be holding their own - though regrettably they have not been granted protection, and are still able to be subjected to the hunter's gun. If these animals *are* to survive, they need the protection to establish a stable population, which could be possible considering the number taken by hunters early in their introduction. The area in which they live is heavy bush, and this has impeded visual sightings, but the evidence is there that this species holds on in their remote piece of Fiordland.

Kawekaweau - The Giant New Zealand Gecko

Hopolodactylus delcourti. In 1986, in the basement of the Marseille Museum of Natural History, a primitive type specimen of a large gecko was found stuffed, unlabelled, and lacking any record of its collection date, or its point of origin. Also, judging from the look of the specimen, it must have been there for over a century.

Once examination of this specimen was performed, it was found to be a species unknown to science. The type specimen consisted of a single skin and partial skeleton - the skin was stuffed and mounted on a wooden plank.

Duvaucel's gecko, Hoplodactylus duvauceli. *Length, 10 in.*

It measured 370mm from snout to vent, and had a total length of 622mm. This meant that this unique specimen of gecko was 54% larger than the largest living Gecko - the New Caledonian giant forest gecko (*Rhacodactylus leachianus*).

Closer examination showed that this gecko was similar to the brown forest gecko (*Hopolodactylus spp.*) of New Zealand (Bauer 1986). This seemed strange, as no geckos that big were present in New Zealand, and the only thing close to that size was the Kawekaweau of Maori folklore. Could this actually be a representative of that mythical animal that had supposedly not been seen for over 100 years?

The Maori said the Kawekaweau represented the souls of dead ancestors of anyone who saw it, and sighting one was an omen - it was that person's time to join them. Thus, they were largely avoided and left alone by the Maori. It was described as being about 2 feet long and as thick as a man's wrist. They were said to be have a brown basal colouration with a red, or sometimes green, striped pattern.

It was found in the forests of Totara (*Agathis australis*) and Kauri (*Metrosideros robusta*) under the bark of these trees. This is typical gecko behaviour - they are believed to be

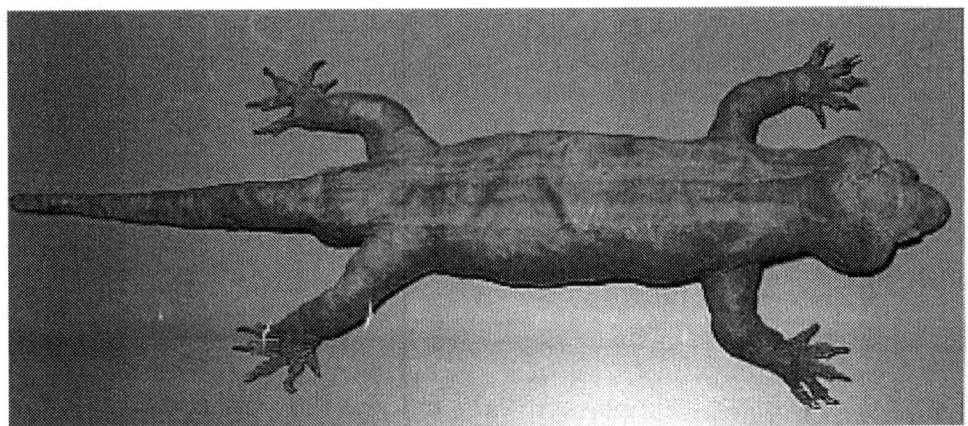

semi-arboreal. Other reports indicated that it inhabited caves and rock piles, and was said to be arboreal and semi-aquatic. The large eyes of the type specimen, together with the bark-hiding characteristic seem to indicate a nocturnal species.

It was supposed to have fed on fruit, nectar, large insects, and small invertebrates, including other lizards, and, no doubt, not excluding its own young. From the size of this lizard, it may also be assumed that it may have included birds' eggs in its diet - with New Zealand's ample avian fauna eggs would not have been in short supply, and would have been a viable food source for this species.

This animal would have been an important predator in its ecosystem.

This gecko seems to have been prevalent in the North Island, and osteological evidence seems to suggest that *H. delcourti*, or a similar large gecko may have been once prevalent in the Otago region of the South Island as well. Bones discovered in the Eamscleugh Cave, in Central Otago, were found in association with rat bones, which indicated they were of quite recent origin.(Hutton 1875).

Captain Cook was the first one to record Maori tales of giant lizards in New Zealand in 1777, including tales of the Kawekaweau. Later European descriptions make vague mention of a pair of these lizards that were captured and kept.

Regrettably, one was eaten by a cat, and the other escaped and was not sighted again. The last known specimen seen alive was caught by a Maori chief of the Urewera Tribe in 1870, while it was hiding underneath some bark. In that same year, a dead large gecko was washed down the Waima creek, from the Waoku Plateau area.

In 1986, when the discovery of the type specimen was made, rumour abounded of a population of these lizards near Rotorua. This rumour was, however, never substantiated.

In April 1990, several sightings of a large gecko were reported in the daily press in New Zealand.

Most of these recent sightings were centred around the Gisborne, the Waipoua Forest, and Waoku Plateau. Some were investigated by local Herpetologists, and though many were dismissed as misidentifications, a few could not be explained as easily.

It would be quite easy for a large species of gecko to still exist in some of the isolated areas of the country, as new species are still being discovered.

Two smaller new species of *Hopolodactylus* were discovered as recently as 1981, and 1984, so there is hope that the larger *delcourti* may turn up. Finding it, however, will be no easy task as its nocturnal habits, possible small population density, isolated locations, and the fact that it hides during the day will make it very hard to find.

After its discovery, the origin of the lizard remained a mystery for quite a while, and some speculated the type specimen may have originated from the French Island Territory of New Caledonia. This would explain it being in a French Museum, and if this were true, it would mean re-evaluating the distribution of the *Hopolodactylus* genus.

The evidence for this was based on the fact that no other New Zealand lizards were at the Museum - most of the specimens at the Museum originated from the Pacific Islands. Regrettably no documentation existed for any specimens donated to the Museum between 1833 and 1869, and it can, therefore, be assumed that the type specimen may have entered the Museum during this period.

If any of these animals still exist, they are at very high risk. Habitat destruction, predation by cats, weasels, and stoats, together with low breeding densities could cause this species, if it still exists, to join those species already extinct and only known from a single type specimen.

Moa- NZ's Big Bird

No animal has had more speculation heaped upon it than the moa. These birds evolved around 85 million years ago when New Zealand separated from Gondwanaland. They belonged to a group known as Ratites; the family includes ostriches, emus and the extinct elephant bird of Madagascar.

Once New Zealand separated from Gondwanaland and became isolated, it enabled the moa to survive in a relatively predator-free environment. This would have been filled with avian fauna, and the only mammals present were three species of small bats. Recently discovered fossil evidence, however, seems to indicate that there were species of both python and crocodile, which may have included moa on the menu.

A04MEH Alamy Images

This isolation enabled these birds to obtain a wide variety of sizes, from small turkey-sized species such as *Euryapteryx curtus*, to the gigantic *Dinoris giganteus* which stood 4 meters tall, and weighed in at 250 kilograms.

All moa species were herbivorous browsers, and ranged in habitat from the alpine regions to the coastal areas.

The main mystery surrounding this bird is: when did it become extinct? Or is it, in fact, extinct? It has been implied that the moa was already on the decline before the human colonization commenced; evidence from Maori butchering and cooking sites show that there must have been an abundance of birds, judging by the wasteful practices engaged in the butchering process.

Meat from the thighs was primarily used, and the remainder of the carcass was left to perish. This is rather surprising as the Maori of old were very conservation conscious to ensure resources were not overtaxed.

This was accomplished by oral tradition and folklore; seasons set out for hunting different species guaranteed a recovery period. Even the name moa is somewhat of a conundrum as it does not appear among the oral traditional legends of the Maori - the term widely used to make reference to this bird was Tarepo.

It seems generally accepted that the large species were supposed to have been extinct by the early 1800s, and *if* not already extinct by then, they had certainly become extremely scarce. However, it is plausible that some of the smaller species, neglected as a food source because of their bigger relatives, may have persisted longer.

The mid to late 1800s produced many reports of large birds witnessed in isolated areas of bush; this was an era of exploration, and regions were being opened up for settlement.
Many reports focused on the South Island, as gold prospectors and surveyors pushed into the isolated interior areas.

One of the most curious reports of this period includes a confrontation between a sheep dog and a moa. The moa turned on the dog after being harassed, and once the dog backed off, the moa was witnessed to bob its head up and down in the direction of the dog, in what seemed to be a possible threat posture.

In both 1931 and 1960, came further reports of large birds in the bush of the South Island, and in 1989 a pair of birds were observed by trampers, once more in the South Island.

All accounts seemed to involve the large species of moa.

In 1990, there were several sightings of large birds, in the Arthur's Pass district, and tracks were found on two occasions. The most recent sighting caught world media attention, when, on 20th January 1993, three companions - Paddy Freaney, Sam Waby and Rochelle Rafferty - were tramping the Craigieburn Range area.

While Mr. Waby paused at a secluded stream for a drink, Paddy Freaney's attention was drawn to a large bird that was watching them nearby. Freaney drew the attention of his associates to the bird, which then panicked and fled. He chased the bird, with camera in hand, and at approximately 35 meters got the now famous photo of the bird. He further discovered, and photographed - after losing sight of it - wet bird footprints on a rock.

These pictures were shown to a Department of Conservation Officer, who expressed the opinion that the bird seemed very much like *Megapteryx didinus*, a sub-alpine species of moa known to have populated the South Island.

Computer analysis was performed on the photo by Canterbury University, and specialists there expressed the view that the photograph was of a genuine large bird, and not some prop staged to look like one. Deer, and other four-footed animals, were further ruled out.

To add to the weight of evidence in support of the photograph being genuine, in the following year - 1994 - a physician was tramping in exactly the same area the snapshot was taken, when he came across browsing damage that was consistent with what is known of moa feeding habits.

In light of the corroborating evidence, the Department of Conservation made no attempt to follow up on what certainly would have been the find of the century.

Is the Moa extinct?

Perhaps, in some of New Zealand's remote areas - and there are still a few - the species may still hold a tenuous grasp on existence. The sad fact is that if it were to be discovered, how simple it would be to conclude what was started in those Maori middens hundreds of years ago.

Waitoreke

Tales of strange otter-like creatures have come from the South Island of New Zealand, and persist to the present day. What is this strange creature? Is it an undiscovered remnant mammalian species, or a stranded and forgotten link to New Zealand's past?

New Zealand is believed to have had no mammals when it separated from Gondwana, apart from three species of bats. Historical evidence, however, states that either an otter, or beaver-like animal, may also have been resident here, or been stranded long ago. The first accurate account we receive of this animal comes from the logbook of Captain Cook.

When he entered Pickersgill Harbor in 1773, aboard the *Resolution*, he reports in his journal the sighting of a four-legged, cat-sized animal, with short legs and tawny-coloured fur. One of the mariners, who also viewed the animal, was of the opinion that it had a distinct jackal-like semblance, but with a bushy tail. When Cook queried his naturalists about this beast, which they had not observed themselves, they proposed it may have been some type of fox, or more likely, one of the vessel's cats that had somehow managed to get ashore.

Another possibility, of which the naturalists and, indeed, Cook may not have been aware, is that Cook and his crewmen may have witnessed one of the native dogs (Kuri). This species of dog is now extinct, but was described as having shorter legs than the average dog, and a very bushy tail. As natives were residing in the Pickersgill area at the time, it may have been one of their dogs that Cook had seen, and, being unfamiliar with the species, it is possible that Cook misidentified the animal as a new species of mammal native to New Zealand shores.

The Kuri was, in fact, not a native of New Zealand, but arrived with the Maori when colonization commenced.

The 1840s brought sightings of a beaver-like creature in the Lake Hawera District of the South Island - this animal apparently built a dam very similar in style to European beavers.

In a letter to his father, Walter Mandel - the son of naturalist Gideon Mandel - reported an animal the natives called a Kaureke. This was a quadruped that was the size of a cat, short-legged, and had a bushy tail. It also apparently laid eggs. This animal was greatly prized by the natives and kept as pets.

Wanting desperately to procure a hide of one of these animals, Mandel offered a reward for one of these creatures - dead or alive. Regrettably, he retained his money, as a team of Maori went up into the bush to capture one, but returned empty-handed.

Von Haast wrote to fellow naturalist Hochstetter in 1862 that he was thoroughly convinced that animals similar to otters were populating the South Island waterways. He had often come across tracks of a web-footed mammalian animal along the mud banks of the Ashburton River – however, the animal responsible remained elusive.

In 1861, a Christchurch newspaper published another otter account, as the animal was pursued by a dog owned by Captain McMillan. The animal retreated to the water and was lost from sight as it submerged. In 1880, one of these otters was shot in the Canterbury district and was claimed to have been eaten by a local Maori fishing expedition. In 1900, stories started to come in from the Milford Track area of otter-like animals that were being encountered quite regularly there.

Lake Te Anau became a hotspot for sightings as people began to encounter otters around the lake regularly. Strange reports of an otter-like animal surfaced in 1939, as some naturalists in the Waiau area sighted an animal they were unfamiliar with, despite their considerable knowledge of the local fauna.

After these encounters, sightings trailed off until 1968, when an otter-like animal was witnessed to leave the Stevenburn Stream, near the Whakea River in the Southland District. The animal checked the area as it emerged from the stream, then signalled to three other semi-adult individuals, which followed her up the bank and disappeared into the undergrowth.

The year 1971 brought another report - this time in the Hollyford River area. A hunter heard splashing as he neared the river, waiting for his comrades to collect him by boat, and he saw an animal very similar in form to an otter scaling the river bank, and sliding back down again as if playing. He witnessed this performance for about 15 minutes before the animal left.

From the accounts, we can deduce the New Zealand otter is smaller than its European counterpart, as some of the witnesses had seen the European species in the wild. A length of about 100 cm is the largest individual described.

Their fur is described as ranging in colour from dark brown to sepia, in most instances with a lighter underbelly. Tracks found, including those by Von Haast, often show webbing between the digits, basically similar to those left by the European otter (*Lutra lutra*). There is, however, another candidate for the mysterious Waitoreke - the size and description of this animal closely matches that of *L. lutra barang*, an Indonesian species of Otter also known as a "Simung". The only discrepancy is the colour of the fur, but this species seems a likely candidate as it is domesticated in Indonesia, and employed for fishing. If this is the case, it may be a subspecies that has developed due to it isolation in New Zealand.

If it is a remnant population, it may well be extinct, or by now, nearly so.

If not extinct already, its stronghold appears to be Western Otago and Southland, with low population densities further found in the Canterbury foothills between the Waimakariri and Opuha Rivers, remote areas of southern Nelson, South Westland, and the Catlins District of south eastern Otago. However, habitat destruction in these regions would have forced them into unfrequented areas, and the principal hope of encountering this species rests in areas that even today remain virtually unexplored in Fiordland, Otago and Southland.

How did these Asian otters get here to start with?

It is possible, with the Tamil artefacts found in this country, as in the case of Colenso's Tamil Bell dated to 1500 A.D, and other jade artefacts, and – undoubtedly - the new Chinese map showing that the Chinese certainly knew of New Zealand (and perhaps even visited here before the appearance of the Maori) that the animals may have arrived that way.

These otter-like animals may be descendants of animals stranded here centuries ago. If this is, indeed, what occurred, they would not have been abundant to start with, and if still surviving with the colonization of man, they may have relocated further into the remoter areas of the South Island to evade human contact. It further seems plausible that these animals were stranded in the lower region of the South Island, as there are no reports of them in the North Island.

Greg Stubbings, a man who has carried out a lot of research in this country into the Waitoreke's possible existence, recently staged an expedition to the areas of the South Island where the animal had been witnessed in the past, to try and find some evidence of the animal's existence.

Despite substantial personal investment in trip cameras, and other equipment, no sign or sighting of the animal was forthcoming. He did, however, state that many of the regions where the Waitoreke was formerly seen were now extremely urbanised.

Until some physical proof of this animal comes to light, its existence will always be another enigma in the history of New Zealand. We can only hope that if it does survive, it is discovered before it does not become too late, and the Waitoreke becomes another extinct species we only discern from rumour and speculation.

New Zealand Panther

Does a group of large mysterious cats roam the wilds of Canterbury?

A relatively unknown New Zealand cryptid is the South Island panther. Sightings of these cats are confined mainly to the Canterbury area, and have led to the suspicion that the animal, or animals, is an escapee from a private collection. Sightings began in 1996, when a woman, who was mountain biking in the Twizel area, saw a large black cat - about the size of a Labrador. The cat was seen at a distance of about 30 metres.

MAF were notified of the sighting, and it was explained away as a misidentification of a large feral cat. However, the cat problem was to come back and haunt them when, in August 1998, another large cat, resembling a mountain lion, was observed in the Dunstan Ranges near Cromwell. Once again, the animal was described as the size of a Labrador and having a dark orange/mustard coloured pelt.

MAF dismissed it once again as another feral cat sighting.

At the same time these sightings were coming in, people in the Mataura area were seeing an animal similar in appearance to a bobcat.

In July 1999 came another sighting of a large black panther - this time ,however, in the Mackenzie Country. A report also came in from a Pest Destruction Officer from Banks Peninsular.

Also in July, a mountain lion was photographed crossing a paddock near Omarama, and once again government officials investigated the sightings. Again, the conclusion reached was another case of mistaken identity involving the giant feral cats of the area, which they assured people could reach a weight of up to 14 kg in the wild.

During December of 1999 occurred sightings of what has come to be known as the "Moeraki Mountain Lion". Canadian tourists saw this cat as it sunned itself on rocks near Moeraki, South of Omaru. It was described as being distinctly mountain lion-like, which these tourists had themselves seen in their native habitat - about 3 metres long and golden coloured. The cat, on being spotted, sauntered from the rocks and disappeared from view.

The tourists were met with some ridicule. The sighting was, however, given some serious consideration by a local restauranteur, who offered a reward for conclusive proof of the creature's existence. None was ever forthcoming.

Another cat to gain fame, and a name, was a large mountain lion-like animal seen in the Lindis Pass area in 1999. It was hiding in the undergrowth, and photographed by a pair of British tourists, and came to be known as the "Lindis Lion".

In early 2001 the Ashford Black cat was once again seen in the Bushside area of Ashford Forest. The winter of 2001 brought renewed sightings on a farm in the Winterslow area of the Ashford Forest. It was another sighting of the big black cat, this time seen in a deer enclosure at twilight.

A similar animal was also seen in the Anama area, and made this area its home for the next two years, creating sporadic sightings until 2003. A black cat was also sighted in the Mayfield area, near Ashburton, in October 2001, when the Fairlington area also became the home of a large black cat.

Another sighting of the mystery cat came from the Ashford forest area in 2003 - this time, however, the owners of the property where the cat was sighted had noted strange behaviour among the stock on their property at the time the cat was sighted. November, once again, saw a team of investigators, together with an Orana Wildlife Park cat expert, descend on the area and look for signs of the big cat's presence, but nothing conclusive was found.

October 2003 also brought renewed sightings of the Fairlington Cat, which was seen lurking behind a fence near the stockyards of the PPCS Meatworks.

A new sighting of another large black panther came in 2005, when it was seen and photographed at Lake Clearwater, in the hills which overlook the Clearwater settlement. May 2005 produced yet another mountain lion sighting, this time in Queenstown, by an Australian tourist. The cat, described as being the size of a Golden Retriever dog was seen in some scrub near the Heritage Hotel, but was reported as moving and walking like a cat.

With the descriptions above, there are three likely candidates for the species of cat found in the South Island.

Each has a melanistic (black) version - the first and most likely for the black cat sightings is the black panther, phantera, melanistic leopards. *Panthera pardus* are also not unheard of, and originate from South East Asia, and the last candidate is, of course, the American mountain lion (*Felis concolor*).

The diverse area of the sightings is not beyond the reach of any of these cats, as they can easily cover 30 km per day, and Canterbury's climate would be well suited to any of these species. These animals would easily be able to sustain themselves on Canterbury's ample rabbit and opossum population.

Judging from the sightings, more than one cat being involved is not an impossibility, and the possibility of breeding occurring should not be overlooked.

Why haven't they been found?

Cats such as these have remained elusive in the British Isles and continue to do so, but unlike the British cats, we actually have a starting date for the phenomena of when the cats came to be in Canterbury - it all started in 1996.

As many people are scared of the ridicule they will face in reporting sightings of these creatures, and the many other cryptozoological animals that inhabit our country, I have no doubt that more people have seen these creatures, but are unwilling to come forward.

To me, this is a very narrow view, as every New Zealander has the right to know and be proud of our un-natural history.

Yes, I can understand reservations in some cases, as there are those people who would be more than willing to try and shoot these animals to bring back proof, if only for the fame and the monetary gain it would bring. I, too, can see the governmental departments that receive such reports, cringe at the amount of money that may have to be spent on expeditions that may end up as a waste of time and resources, or even their concern over panic as in the case of the big cat sightings. However, the New Zealand public also has a right to know that these things exist, and to be able to make up their own minds.

Populations of moehau, waitoreke, and even perhaps moa may exist in areas known only to a few people too scared of the response they may get to say anything. Or are there those that do know these animals exist, but would much prefer they remain a mystery to the mainstream public, preferring knowledge of their existence to remain shared among a select few?

Manbeasts Of New Zealand

The Maori settled in New Zealand around 1100 AD, and from the early days of colonisation they spoke of mighty ape-men who resided in the forested areas. These areas they dared not venture into, as these hairy beasts were known to tear men apart – they had huge lacerating fingernails, and fed on human flesh. They were known by many names: Moehau, Maero, Matau, Tuuhourangi, Taongina, and Rapuwai. The Moehau were described by the Maori as being *"Terrible creatures, half man, half animal"*, with a very aggressive temperament, and they were only too happy to massacre, and eat anyone that strayed into their domain.

Early encounters often talk of these creatures exhibiting aggression and throwing rocks to frighten people off. It was these creatures, largely found in the Coromandel Ranges, that were thought to be responsible for the find of a headless, partially devoured body of a prospector in the Martha Mine region in 1882. Later, further up in the foothills, the corpse of a woman was found. It was discovered that she had been dragged from the shack in which she lived, while the remainder of her family were away, and her neck had been snapped.

On the topic of aggressive behaviour, Toanginas were greatly feared by the population of the lower Wanganui River, as they were said to viciously attack any fishermen who strayed into their territory. This vicious behaviour, however, seems to have abated in more modern encounters, as the beasts in most instances flee on sight of humans.

Rapuwai are believed, from legend, to be able to crush any strong Maori warrior with ease, employing their large powerful hands. They are said to be tool-producing beasts using wood and stone, and the articles crafted are said to resemble those produced by *Homo erectus* hominids.

Moehau are depicted as being as tall as a man, completely covered in hair, with marginally ape-like facial features. The primary difference from human appearance are the extremely long fingers, tipped with sharp talons, capable of tearing apart the toughest prey. They are often described as "Manimals".

It is possible that, if these man-beasts existed prehistorically, they would have been more than capable of bringing down the largest of moa (*Dinoris giganteus*). The large talons spoken of, seem to designate this creature's predatory nature. However, large talons are also found elsewhere in the animal kingdom in animals that rip open rotten logs to acquire nourishment, and considering the indigenous Maori used to eat the large nutritious huhu grubs, it is not impossible that this beast may also be insectivorous.

- CFZ YEARBOOK 2008 -

Artist's impression of Coromandel man

Matau giants are described as being ape like but 3m tall.

The Rapuwai are gigantic, slow clumsy creatures that are of a strong muscular stature. These creatures can be categorised as follows: those that are the stature of an ordinary human - the Moehau and Maero - and those that are of giant stature - the Matau, Tuuhourangi, Matau Giants and the Rapuwai.

Many areas of New Zealand are named for these great hairy man beasts: Moehau Mountain, where they are believed to reside and people are cautioned against going up there is one such place.

The Moehau are thought to populate both Mount Tongariro and Ruapehu, the Karangahake Gorge, Coromandel Ranges, Martha Mine Region, Waikaremoana – in the Urewera Ranges, the Heaphy River of the Northwest Nelson State Forest Park, Kaikoura Mountains, Fiordland National Park and are believed to be very common in the Haasts Pass area, particularly around the Haast River.

The Matau Giants inhabit Lake Wakatipu in Central Otago. Toanginas are found in the lower reaches of the Waikato River. Maero are encountered in bush country throughout both the North and South Islands. Rapuwai are said to inhabit the Marlborough Distract, and the Milford Sounds area. There is another, as yet unidentified, type of man beast that lives in the Cameron Mountains in the South West of the South Island.

Why have no bodies of these beasts ever been found?

I think that is readily explained by the fact that these 'Manimals' may have a conception of death, and bury their deceased. If this is the case, the New Zealand bush does not give up its dead readily, the chances of finding a buried carcass, unless you know precisely where to look, are reasonably slim, especially in some of the more inaccessible areas.

It is possible that some of these man-beasts may still exist in the more remote isolated areas of bush throughout both Islands. With habitat destruction and human encroachment, this species, if it survives, must unquestionably be on the brink of extinction, or may be already extinct. It appears the last bastion of the Moehau is in the Coromandel Ranges, where accounts seem to indicate they resided in their greatest population density.

Footprints are, in most instances, the main evidence of these creatures. In 1903, footprints larger than a man's were found in the Karangahake Gorge in Coromandel. In 1971, a trail of footprints, similar to a man's - though extended in appearance - was located on snow-covered ground by a Park Ranger, and led into a zone of bush on a hillside. In 1983, a deer hunter chanced upon man-like footprints that could have been no more than an hour old, along a riverbank in the Heaphy River area. In 1991, campers in the Cameron Mountains of the South Island elected to abandon their camp, after finding unusually large man-beast prints near where they were camping.

In 1970, another party of campers had to abandon their camp as a 2m tall man beast assailed them screaming loudly, and hurled rocks at the camp.

In 1972, a hunter in the Coromandel ranges watched a naked, hairy man beast about 2m tall, work its way through the scrub on the other side of a gully, and upon reaching the place the creature had been transversing, footprints were found.

An undated sighting surfaced from the *Sunday News* stating that an occupier of the Lake Mahinapua Pub had been having his vegetable garden raided regularly. The offender was exposed when a man beast was witnessed dashing to the protection of the bush, with an arm load of silverbeet.

Regrettably, some of the Coromandel sightings may have been of an escaped gorilla. A vessel anchored near Wai Aro in 1924 had a pet Gorilla as a mascot, and the animal managed to slip away from the ship, and was never captured. However, sightings of the beast provoked a few startled reports from the region.

SINGING MICE

Some years ago I wrote a major research paper for the now defunct *Fortean Studies*. It was an exhaustive review of the literature on singing and dancing mice. The story behind my researches is told in this excerpt from my book *Monster Hunter* (2004):

> I first discovered the story in the Westcountry Studies Library. This is a boon for all researchers living in the Exeter area, and I have used it again and again. As well as an unparalleled collection of books, journals and documents on the subject of the four counties of the West Country it also contains a complete collection of the two local newspapers going back well into the 19th century. At the time I was researching a book, that I later decided not to write, about the mystery animals of Devon and Cornwall, and was on a ceaseless trawl through newspaper archives in search of zoological anomalies.
>
> An item in the 43rd report of the Scientific Memoranda section of the *Transactions of the Devonshire Association* caught my eye. The headline read *"Singing Mouse"*, and it quoted an item from the *Western Morning News* of 2 February 1937, which referred to:
>
> *"Mickey, the singing mouse caught by Mrs A. Eddey of Trafalgar-place, Stoke, Devonport.*
>
> *Prof. Crews, of Edinburgh, wishes to investigate the vocal mouse in*

the interests of science, but Mrs Eddey's primary wish is that it should broadcast..."

This was an opening paragraph worthy of anyone's attention. A singing mouse whose owner had showbusiness aspirations was a beast out of the pages of one of the *Dr. Dolittle* books. Mrs Eddey herself seemed almost prosaically matter-of-fact about the whole affair:

"It is true I have a lovely little singing mouse which I caught on the morning of January 10th. It sings like a canary. It is an ordinary house mouse, very small, and its little body seems to vibrate with music. It first came to my bedroom at 12. 2 [sic] A.M. on New Year's morning and it has sometimes sung the whole night. Even when trapped it did not stop chirping to me. I am sorry I can tell you nothing more, only that it is quite tame..."

The headline in the *Western Morning News* read *"Mouse that has won fame"*, which seems eminently appropriate for such a peculiar story. George Doe, the Recorder of Scientific Memoranda for the Devonshire Association, reported that 'Mickey' was again in the news, when the *Western Morning News* of 12 March 1937 reported that Mrs Eddey's ambitions for her pet had been fulfilled:

"Apparently pleased with the success of his broadcast audition on Wednesday, Mickey, the Devonport singing mouse, kept his owner awake all night with his celebration tunes.

If all goes well, Mickey will soon be issuing a tuneful challenge across the ether to Minnie, his American rival. On Wednesday, Miss Mildred Bontwood, of the National Broadcasting Company, of America, travelled to Plymouth especially to see Mickey, having previously wired his owner Mrs A. Eddey, of Trafalgar-place, that she was bent on seeing him.

Mickey was put into his cage and taken around to the Plymouth Station of the British Broadcasting Corporation, and as a result of the audition Mrs Eddey has a contract to take the mouse to London and it is expected that Mickey will broadcast from there to the United States."

By the time I had finished reading this extraordinary series of articles I was almost in tears of laughter and was in imminent danger of being asked to leave the library. It all seemed too surreal, even for inclusion in a universe which experience has taught me and my colleagues is often totally absurd. The fact that the whole report had been compiled 54 years before by someone whose surname is commonly used by American policeman as a designation for unidentified corpses made the whole affair seem even more bizarre.

I decided that before proceeding with the affair the sources should be checked. I had the easy job. I checked the relevant issues of the *Western Morning News* and found that Mr Doe had indeed quoted the original press reports correctly. The BBC still had a copy of the recording, but unfortunately the documentation that went with it had gone astray. Alison, however, had the unenviable job of trying to find out something about Mrs Eddey.

Even if she had been a relatively young woman at the time, by 1991 Mrs Eddey would have been in her seventies. The odds were that she was dead. Trafalgar Place in Stoke didn't exist any more, and even if she were still alive, tracking her down seemed as if it was going to be an impossible task.

We were right. It was!

In 1991 there were 26 telephone numbers under the name of 'Eddey' in the Plymouth telephone book with Devonport addresses. Alison telephoned them all. Unfortunately none of them had heard of either Mrs Eddey or her talented rodent. Some people were helpful and polite, others abusive; none had any information that was actually any use in our quest.

There were several hundred people with the same surname living in other parts of Devon so we did what any sensible Fortean researchers would have done under the same circumstances. We gave up!

Instead of continuing the search for Mrs Eddey or one of her close relatives, we went back to searching the newspapers for more stories about this remarkable rodent, and we were not disappointed.

Back in 1937 the plot had thickened, as by April 'Mickey' had another British rival.

The headline in *The Times* on 22 April 1937 was:

"BBC Tests the Singing Mouse - (like a nightingale)"

The story read:

"Another singing mouse has been found - this time in Wales. It will be brought to the microphone on May 8th for a national broadcast, which will be relayed to the United States. The mouse, 'Chrissie', owned by Mr Gale of West Cross, Mumbles, Swansea, was given a test at the Swansea studio of the BBC and an official told a press representative that she sang 'like a nightingale'. For the test the mouse was held up to the microphone in a bottle.

The National Broadcasting Corporation of America has challenged any country to produce a singing mouse to beat the singing mice of Illinois. Mr Gale discovered the singing ability of his mouse last Christmas. Soon afterwards 'Chrissie' was missing. She was found hiding inside the piano. Before this it was claimed that the only singing mouse in Britain was that owned by a Devonport woman, which has also been tested for broadcasting. "

The Illinois mice had made their media debut the year before on a Detroit radio station when on 15th December they broadcast a recording of "Minnie the Singing Mouse" to mixed responses from the audience. *"Some people thought that she sang like a robin, others compared her to a tone-deaf canary. The trouble with Minnie was getting her started, but once she opened her mouth she wouldn't stop."*

Three days later, according to Young, another Illinois mouse, named 'Mickey' and billed as her 'co-star', made his debut with a less than impressive performance. He, apparently, got his feet wet while drinking water from a fruit jar, and refused to sing. After half a minute's silence an announcer told listeners that Mickey usually began a recital with a soft whirring trill rising into a crescendo, followed by a two-note jump, tapering to a diminuendo. No-one believed him.

Writing in a book called Secrets of the Natural World, *my friend and colleague, the British cryptozoologist Dr Karl P. N. Shuker, also described 'Minnie' and her career, and told how "…. in May 1937, a transatlantic radio contest for singing mice was staged, featuring rodent songsters from as far apart as London, Illinois and Toronto. "*

The first English entry was a duet between Mickey from Devonport and Chrissie, a Welsh mouse. They piped quite brightly, but no-one could tell them apart. America's Minnie (from Illinois) ran round and round and refused to open her mouth. Mickey (also from Illinois) performed like a trouper, his top notes were described as being comparable with the greatest Italian tenors of the day.

Next came Johnny of Toronto, billed as the Toronto Tornado. He never had a chance; tens of thousands of radio listeners heard cries of 'Miaow! Miaow!', followed by a solemn announcement that the Tornado's career had ended.

It was back to London for another 'Mickey', but listeners mistook him for a leaky tap in the radio studio. At the end of the contest the sponsors (Canadian, American and British Broadcasting Corporations) announced the winners would take part in an international mouse opera broadcast live. Over half a century later we are still waiting for the grand results of the competition. Both NBC and the BBC have ignored my requests for the winners' names to be publicised, and to our knowledge the mouse 'grand opera' so eagerly awaited by Dr Dolittle and his friends has never been staged.

Young treated the whole affair in such a light-hearted matter, that, although there is enough corroboration to confirm that the event actually took place, the true details are obscure. We have succeeded in part of our original aim, however. 'Mickey' the singing mouse of Devonport definitely existed. He was not an unlikely hoax dreamed up by George M. Doe. There is even a picture of him, but there do not seem to be any records of his life after May 1937. Like so many semi-legendary historical characters, little is known of him apart from his brief flirtation with fame for five months in 1937. Mice don't as a rule live very long, and so we can reasonably suppose that he has passed over to pastures new.

The same can be said about the other mice in the story. Essentially the main part of our research was over. We were, however, still intrigued. Why had so many singing mice turned up between December 1936 and May 1937? What made them sing? Was there a historical precedent? Were there singing mice around today? If so, where could we get one?

I have to admit that my motivation at this point was not merely the furthering of the sum total of human knowledge. I have always had a strong and irrational dislike of Walt Disney. In the early 1980s there was an inept American punk rock ensemble called *Bomb Disneyland* and although the music was terrible I bought all their records because, unusually amongst the practitioners of American punk rock, here was a sentiment with which I could sympathise. I was particularly incensed at what Disney Studios had done to great classics of English literature like *The Jungle Book* and *Peter Pan* and it would have amused me greatly to have proved that the whole Disney Empire had been based on an idea stolen from as peculiar rodent living in a suburb of Plymouth.

So, fuelled by a heady mix of righteous indignation and a surreal sense of humour, and ignoring the protestations of my wife I continued my researches. I discovered a number of letters to *The Times* discussing singing mice which were published during the 1930s and over a few months I amassed what I believe to be the largest archive of material on the subject to exist in the world.

It seemed that the phenomenon of 'singing' mice is quite a well-known one. As with so many things the phenomenon captured the interest of a number of Victorian writers and naturalists who discussed the matter at length. It seems that, despite our initial surprise in the Westcountry Studies Library, singing mice were a well-known phenomenon amongst out forefathers, and that it is only the effete researchers of Forteana and zoology at the end of the 20th century who had not heard of them. It was a little comforting to find out that with three exceptions, Dr Karl Shuker and my friends Clinton Keeling, the veteran British Zoologist and maverick zoological researcher Richard Muirhead (and they know all sorts of ridiculous things), no-one we would talk to during the rest of our investigations had heard of them either!

The 'craze' for singing mice during 1936 and 1937 led to much information becoming available. In *The Times* of April 22nd it was stated that a singing mouse had been found in Wales and that it was to broadcast on May 8th. A letter in the same issue informed the British populace that according to Red Indian mythology, 'Mish-a-boh-quas' the singing mouse always comes to tell of war. It may sing at other times but not to the same extent. The author of this letter cited "Ernest Thompson Seton's wonderful book *'Rolf in the Woods'*."

This was just too much for me to deal with. What had originally been a mildly amusing piece of research into an obscure item of Devonshire Zoology now seemed to have analogues all over the world. A good friend and colleague of mine is a bloke called Tom Anderson. He lives in Scotland and it is a mark of the peculiar nature of the technological society in which we live that we have become firm friends by letter, e-mail and telephone conversation without ever having actually met face to face.

Besides being a highly amusing raconteur of tall tales and a collector of strange Scottish stories, Tom is an expert on matters appertaining to the Native American peoples, did some research for us but was unable to unearth any solid facts. He wrote to me:

"Despite strenuous efforts and consulting about thirty books on folk customs, anthropology and totemic realisation and ceremonial, I can find no reference to singing mice Their name sounds Algonquian, which limits them geographically but it doesn't appear anywhere, even as a sub-clan of a linguistic family such as kiowan, Siouan, Athabascan, etc.

Mice are an unusual tribal choice, either as totems, guardians or emblems. Not having the power of the Thunderbird to control the elements for the northern tribes, or the storytelling significance of the Coyote or Grandmother Spider further south, it's difficult to imagine its purpose. The nomadic tribes used medicine bundles for personal protection and luck, as you probably know. Feathers, lucky stones, weasel skins, claws, etc., symbolic of speed, running and other desirable traits.

Custer's nemesis, the Oglala, Crazy Horse (Masunka Witco), whose war paint consisted of painting his face blue with white spots, symbolising hail, and wearing a sacred stone behind one ear and a hawk's body in his hair, was the rare combination of a mystic and a war leader"

Tom continued describing Crazy Horse for some paragraphs before returning to the main subject:

".... mice are representative of nothing I'm aware of in Indian culture. Nor can I find evidence of them as design subjects for the Pueblo cultures on pottery, etc. Deer, yes. Mice, no!

He concluded:

I suspect it could be a 'retro myth'. A possibility, maybe even likely, but basically unfounded. It wouldn't be the first as Amerindians are 'hip' right now because of their 'green' lifestyle. I won't bore you with the shortcomings of this theory."

To confuse things further, Arkady Fielder wrote a book called *The River of Singing Fish,* referring to a type of sheat fish in the Ucauali River, which apparently produces a noise like bells clanging. But no mice. Richard Muirhead, a bloke I had actually been at school with back in Hong Kong nearly three decades before eventually managed to track down the passage in the aforementioned book by Seton:

"A few nights later. As they sat by their fire in the cabin, a curious squeaking was heard behind the logs. They had often heard it before, but never as much now. Skookum turned his head on one side, set his ears at forward cock"

At this point, we feel, the reader should be reassured that 'Skookum' is the name of the dog owned by the eponymous hero, Rolf, during his sojourn in the woods. The narrative continues:

"Presently, from a hole 'twixt logs and chimney, there appeared a small, white-breasted mouse. Its nose and ears shivered a little, its black eyes danced in the firelight. It climbed to a higher log, scratched its ribs, then rising on its hind legs, uttered one or two squeaks like they had heard so often, but soon they became louder and continuous:

'Peo, peo, peo, peo, peo, peo, peo, oo,
Tree, tree, tree, tree, trrr,
Turr, turr, turr, tur, tur,
Wee, wee, wee, we'

The little creature was sitting up high on its hind legs, its belly muscles were working, its mouth was gaping as it poured out its music. For fully half a minute this went on, when Skookum made a dash; but the mouse was quick, and it flashed into the safety of its cranny.

Rolf gazed at Quonab inquiringly.

'That is Mish-a-boh-quas, the singing mouse. He always comes to tell of war. In a little while there will be fighting.'"

There are times when I look at my life as if from an outsider's point of view that I think that everyone I know is either massively eccentric or barking mad. Some are both. For Richard was so excited by his discovery of the original 1915 reference to the Native American singing mice as a portent of war that he decided that merely posting me a photocopy of his discovery, or even telephoning me would be a wholly inadequate way to communicate this momentous discovery to me.

Nothing so tame! Despite the fact that the weather was appalling and it was a Bank Holiday weekend he decided to trace (I believe by a mixture of hitch-hiking and railway train), all the way from where he lived at the time in Salisbury to my home in Exeter in order to tell me of his fantastic find in person. Unfortunately he forgot to warn me of the fact. Alison and I were upstairs watching television when he arrived and didn't hear his repeated knockings. Apparently he was standing on my doorstep for several hours singing the Red Indian mouse song a-la Ernest Thompson Seton before giving up and hitch-hiking back to Salisbury, still without having given me the original documents!

Rolf in the Woods is a dreadful book. From the photocopied excerpts eventually sent to us by Richard Muirhead we are exceedingly glad that we didn't have to read more of it than was absolutely necessary. It is also a work of fiction, although the author claimed to have based it on his own experiences. It also appears, although we cannot confirm this, to be the original source for both *The Times, and* the Young references.

We have not been able to find any other pre-1937 references to this legend. In the absence of any further supportive evidence we are forced to conclude that Tom Anderson may well be right and that the story of *Mish-aboh-quas* may well be nothing more than a relatively modern 'refto' myth based on an incident in a (not very good) novel.

Whether or not the Native Americans did have this legend, we can establish that native North American mice do sometimes exhibit this 'musical' ability. So far all the mice which we have discussed which exhibited this 'musical' ability appear to be common house mice *(Mus musculus)*, originally a Middle Eastern species which, commensal with man, has spread to every corner of the globe. Writing in 1871, however, William Hiskey noted:

"The cage had a revolving cylinder or wheel, such as tame squirrels have. In this it would run for many minutes at a time, singing with its utmost strength. This revolving cage, although ample as regards room, was not over three and a half inches long, and two and a half inches wide. "

It could be argued, perhaps, that the sound of a mouse 'singing with its utmost strength', was in fact the sound of a desperately unwell animal wheezing and gasping for breath whilst rushing around in its wheel. I feel, however, that this hypothesis is unlikely, if only because of the unlikeliness of an animal suffering from such a severe and debilitating condition voluntarily taking exercise to this extent. And anyway, Lockwood was a fine naturalist, and it would seem eminently unlikely that a man of his observational powers would have been unable to recognise diseased squeakings when he heard them. As we explored the surprisingly large body of source material about singing mice, it became more and more apparent that whilst it seems entirely likely, nay probable, that many of these 'singing' mice were, indeed, suffering from a debilitating respiratory tract infection, others, probably including 'Hespie', were exhibiting behaviour symptomatic of something else entirely. When an animal is ill, it is usually self-evident, especially to a competent naturalist, and Lockwood, in particular, spent some considerable time testing the 'respiratory infection' hypothesis, before rejecting it out of hand.

The final irony is that if indeed 'singing mice' are suffering from various respiratory tract infections, then the ancient Wiltshire folk legends of the advent of a singing mouse foretelling sickness could be nothing more than literal truth. The story of the 'Black Death', a global pandemic spread by lice living on the (then) ubiquitous black rat, is well known. If there is/was a disease of the respiratory tract which affected both mice and humans (and we should here remember that one of the best rationalisations for vivisection experiments on mice is that they are biologically relatively similar to our own species), then the advent of a diseased mouse could well have been the forerunner of an outbreak of disease amongst the human beings of the neighbourhood.

Unfortunately, it turned out that my main objective in this quest was fruitless and it was not possible to prove that Walt Disney made his fortune from an 'empire' based around the bastardised images of two rodents with debilitating upper respiratory tract infections, but the sheer peculiarity of our quest fascinated me and inspired me to go on stranger and more peculiar investigations into the soft white underbelly of the natural sciences.

We are indebted to Rod Dyke from Washington for sending us copies of two of the most obscure American articles written on the subject of singing mice, which we reproduce in fascimile on the next four pages.

They are seventy or so years old now, and this is the only chance that most readers of this volume will ever get to read the originals.

December 28, 1936

Singing Mouse

Inmates of the Chicago Industrial Home for Children at Woodstock were convinced last fortnight that a canary was loose somewhere in the building. Day after day they heard it chirp and trill. Day after day they searched for it high & low without success. Then one day the school manager heard the piping behind him, turned and beheld its astonishing source—a small, greyish brown mouse.

Captured, placed in a glass jar and named Mickey, the singing mouse became the wonder & delight of school and neighborhood. Even newshawks admitted after an audition that it actually sang. When Assistant Director Robert Bean of the Chicago Zoological Park called, it failed to perform. Nonetheless Director Bean, who had heard of singing mice before, offered $150 for it.

Dr. Wilfred H. Osgood, zoology curator of the Field Museum of Natural History, also said he had heard of singing mice, though he had never seen one. Declared University of Chicago's Dr. Maud Slye, famed cancer experimenter: "I have had 160,000 mice and I never had one that sang. If there is a singing mouse, I am open to conviction."

To the mouse, renamed Minnie after examination by Zooman Bean, came a supreme test one evening last week. Up to a microphone in NBC's Chicago studios stepped the master of ceremonies of the NBC Jamboree to announce "the phenomenon of the century . . . the only mouse in the world who actually sings." Into the studio marched the Industrial School's tall, gaunt Manager Oscar Alva Allred, carrying Minnie in a wire-fronted box. Holding the cage before the microphone, Manager Allred poked a small piece of insulated wire through a hole in the box top, tenderly prodded Minnie's belly. As the visible audience of 400 listened raptly, out over a national network went faint, wavering chirps and trills. It sounded as much like a cricket as like a canary, but that Minnie really sang there was no doubt. After the broadcast a cage was fashioned of glass and cardboard, its bottom strewn with strips of cloth and paper for mousy nesting. Press and newsreel photographers crowded around, snapped perky, self-assured Minnie until midnight. A Chicago hotel matched the Zoo's offer for her. Manager Allred held out for $1,000, hoped to get it from Walt Disney, whose singing mouse escaped few months ago.

How & why mice sing is a scientific mystery. Dr. Slye thought Minnie might have a respiratory condition similar to human râles. In 1932 Zoologist Lee R. Dice of University of Michigan suggested in the Journal of Mammalogy that all mice may sing, but on a pitch too high for the human ear unless the mouse has unusual vocal equipment. In other words, perhaps Minnie was a basso mouse.

Dr. Dice, president of the Michigan Academy of Science, Arts & Letters, secured a singing mouse in 1926, has bred many descendants without producing another real songster. Last spring he reported to the Yale Institute of Human Relations the mouse's superiority to the canary as a musical pet. Observed he: "The musical mouse can be heard only 25 feet, so that the song is less irritating to the nerves and can be escaped easily by moving out of range."

> *Natural History,* February 1937

The Indoor Explorer

By D. R. Barton

THREE SINGING MICE: Radio listeners throughout the country were recently astounded by the announcement, broadcast from a Chicago station, that they were about to hear the vocal efforts of the much publicized singing mouse lately discovered in a children' s industrial home near that city. No listener could have been more nonplussed than your indoor explorer as this strangest of radio debuts, a series of quite musical chirps and trills, reached his incredulous ears.

Having had no previous experience with the vocal achievements of mice other than the usual squeals of fright followed by a scuttling into nearby holes, the writer had supposed that their voices were never more than a means of expressing the most fundamental emotions as briefly as possible. The idea of a mouse radio-entertainer trilling and chirping with professional nonchalance into a microphone gripped your explorer' s imagination, and caused him to murmur: "What wouldn' t I give for a singing mouse story of my own?"

Though not so suddenly answered, it was almost as if a fairy god-mother, hearing the wish, had waved her wand; in fact she must have waved her wand thrice for, a week or so later, the writer was invited to a concert given by, not one, but three singing mice.

The concert took place in, of all places, a New York apartment; but this was not an ordinary apartment—in fact it is probably the only one of its kind.

Tenanted by Mrs. William Le Roy Cahall, a member of the American Museum, it contains a living collection of approximately 300 tropical birds ranging from a 30-year-old large-sized parrot, to a tiny song bird about the size of your thumb. Sleeping at night in cages, all these birds have the complete run of the apartment during their waking hours. A flock of canaries, fifty strong, sweeping into the living room, circling, and flying in close formation out along the hall again is a common every-day sight for Mrs. Cahall' s visitors.

The extraordinary breeds the extraordinary. It was therefore fitting that with all the apartments in this particular building to choose from, the three minstrel mice should have selected Mrs. Cahall' s as their winter residence.

Of course, what really attracted them was a small room which Mrs. Cahall has set aside as storeroom for the large quantities of grain and seed necessary to feed her magnificent bird collection.

One evening, a short time ago, Doctor and Mrs. Cahall heard a soft trilling sound which, to their astonishment, evidently came from this storeroom.

The song resembled that of a female canary.

"One of canaries has slipped in there instead of going to his cage to sleep," concluded Doctor Cahall.

"It doesn't seem possible," replied his wife, "they have never disobeyed me before. When I call them each and every one has always gone to his own cage to sleep, and besides the storeroom door is never opened at bed time."

Fearing that one of the canaries had perhaps hurt its wing and had been unable to get out of the storeroom before it was closed, Mrs. Cahall entered and switched on the light—what she saw sent her running into the living room shouting to her husband to "come and see mice—three of them—that sing like canaries, nibbling at one of the seed sacks." Doctor Cahall laughed. "Impossible," he said. When she returned with her husband the mice, frightened, had disappeared. Doctor Cahall enjoyed his joke—but not for long.

The next evening, at about the same time, the singing was heard again. This time Doctor Cahall saw the mice: a little one with a shrill, melodious "soprano," a medium-sized mouse with (shades of the three bears) a medium-sized voice, and a big fat one with a deep "chirrup."

These mice have continued to provide the Cahall's with entertainment.

They form a trio and harmonize rather well. As they are quite wild and shy, you must crouch in the dark storeroom to hear them. When the light is flashed on they soon run to cover, usually, however, continuing their song in flight. The little soprano is the most proficient of the three, Mrs. Cahall asserts, and is also the most persistent. Sometimes, unaccompanied by her "supporting cast," she will give solos while on a solitary forage in the storeroom.

"I say 'she' because the little mouse seems like a female," explained Mrs. Cahall. "Although she is the tamest of the three, I have not succeeded in getting on sufficiently good terms with her to examine her very closely. But I am making progress. Just the other night I got her to eat a bit of cheese out of my hand. I want her to get accustomed to the smell of my hand and to associate it with food. In this way I hope gradually to tame all three mice and add them to my pet collection."

The three songsters have all the physical characteristics of the common house mouse. They are steel grey in color. When singing they often sit upright on their haunches, place their fore-feet about their heads and twitter in their respective keys very much in the attitude of dogs baying at the moon.

Mrs. Cahall says that she has every reason to believe that these songs are not indications of impaired breathing or of a diseased condition of the vocal chords. Her opinion is that the mice sing because they want to, and that, if the song means anything at all, it means that the mice are enjoying themselves. She bases this assumption on the fact that the songs are usually forthcoming at feeding time.

Just why the mice did sing was indeed puzzling. The writer made the tentative suggestion that they might have learned the habit from the canaries. This, Mrs. Cahall doubted, since the mice never sang in the daytime when the canaries were singing, and could scarcely be expected to wait modestly until their unwitting instructors had gone to roost before piping their soft-toned imitations.

Nor does one find any support of the "imitation theory" among the papers of Prof. Lee R. Dice of Michigan, who has done considerable laboratory experimentation with singing mice. Not only is Prof. Dice skeptical of the idea that singing mice imitate birds but he doubts that any musical sounds influence them very much. Says Prof. Dice: ". . . the captive male [singing mouse] in my possession did not respond in any way to musical sounds of many different sorts produced for his benefit by the phonograph and by various instruments. This does not agree with the observation of Brehm who states that singing mice are stimulated to sing by piano music."

Despite exhaustive research on the subject, Prof. Dice does not make a definite statement as to the cause of the rare occurrence of a musical song in certain otherwise normal house mice.

After examining many suggestions by others and stating the number of alternatives that occur to him, he concludes that there is a possibility that all mice are singing mice. His experiments do not show, however, that this odd vocal condition (or whatever else may he the cause of the song) is an inherited character, but that ordinarily their song is pitched too high for the human ear and that only an occasional mouse with an oddly constructed vocal apparatus gives song that is audible.

Prof. Dice was able to study the heredity factor in minute detail in his breeding laboratory and found no combination of mates that would result in offspring capable of even approaching the vocal qualities of their sire. However, he feels that more research is needed before this matter can be finally settled.

So your explorer is forced to conclude that singing house mice are simply unpredictable. A person who doesn't take the slightest interest in animals, usual or unusual, may wake up some bright morning with a whole family of singing mice twittering cheerfully at the foot of his bed. While someone who would give his eye-teeth to possess one of the rare creatures might spend a lifetime in fruitless search.

Reports of their existence have emanated from every quarter of the globe. In certain countries where song birds are scarce or too expensive for slim purses, singing mice have been kept in cages as songster pets. Travelers in China during the last century told of the value set upon singing mice as pets in that country, and communities in southern France were said to keep them in notable numbers.

However, so far as your explorer knows, no one has succeeded in setting up, side by side, a collection of every type of canary together with a family of singing mice. And if Mrs. Cahall is able to tame her mice and bring about this unprecedented relationship, she will have accomplished something that is not only unique in itself but is of value to scientific knowledge.

Both the Department of Mammals and the Ornithologists of the American Museum would be intensely interested in the possibilities of such a liaison.

If she is successful, it will not be the first time Mrs. Cahall has benefited science. She has been a sort of avian cornucopia for the Central Park Zoo, having contributed over 700 birds of all types for public exhibition.

She has also made a study of bird ailments and has learned to cure many of the diseases she has studied. Although the wife of a physician, she took no special interest in the pathology of birds until such an interest was forced upon her. "I learned to care for sick birds," said Mrs. Cahall, "because I could find no veterinary to heal my pets when they fell ill."

It is curious how the various problems that present themselves in the course of pursuing a hobby grow so great that, to solve them, one's avocation must rapidly become one's vocation.

Your explorer is wondering to what lengths Mrs. Cahall's latest hobby, her singing mice, will lead her. He for one is prepared to await developments with the greatest of interest.

CFZ ANNUAL REPORT 2007

Dear friends,

Counting up on my fingers, I realise – with a start – that this is the fourteenth annual report that I have written for the membership of the CFZ. Every year since 1994, during the first week in December, I sit back and reflect upon the events of the previous twelve months, and then attempt to put them into writing.

This has been a tumultuous year, both for the CFZ and for me. On a personal level, just over two years after we met, Corinna moved in with me full-time in April, and we married on 21st July. This is, therefore, the first annual report that I have ever written as a happily married man. But the advent of Corinna has not just meant that my years of turmoil are over, but she has been faced with the monumental task – almost akin to that of cleaning the Augean Stables – of bringing order to the chaos that was the CFZ administration. And she is doing a magnificent job. We have spreadsheets, accounts, databases, and even a franking machine (although those jolly nice people at the franking machine company have still not given us the facility of printing our logos on outgoing mail). It is an epic task, but – assuming that we survive the forthcoming recession intact – one that has already to bear fruit.

One of the most important results of our newfound efficiency is that, at last, we are beginning to attract corporate sponsorship.

The first of these sponsors was Travis Perkins, who very kindly donated about £800 worth of timber, which has been used in the construction of the museum, and the second is Capcom – one of the world's leading computer game publishers – who very kindly sponsored our November expedition to the South American country of Guyana.

But more of that later.

Work on the museum has been hindered by the bloody awful weather. It is wryly amusing to look back and realise that in April we were told by those in the know, that 2007 was going to be the hottest, and driest, year on record. In fact, it was anything but! We had hoped that the museum was going to be – to a certain extent at least – ready for visitors in time for the Weird Weekend. As June, July, and August were almost unrelentingly horrible weather-wise, practically no work was able to be done, and so the revellers who attended first our wedding, and second – a month later – the Weird Weekend, were confronted by a dilapidated, and rather unsavoury-looking building site. However, the floor is complete, the electric supply has been installed, and the aviary block is 95% complete. There is a hell of a lot more to be done, and it still looks like an unsavoury building site, but we hope that the vast majority of work will be completed by the spring. However, it will be eight months late, at least, and between £5,000 and £8,000 over budget.

However, it will be the only institution of its kind in the UK, and will be open to the public on selected days during 2008. We can announce the first four days of these: the last weekend in June as part of the Open Gardens project, and the Weird Weekend on the third weekend in August. Otherwise, it will be open by appointment only, and - all year round - to all members of the CFZ.

We are also proud to be able to announce our involvement with a major conservation initiative. Chirs Moiser, who has been a member of the CFZ permanent directorate since 1995, has just bought a zoo – *Tropiquaria* at Watchet in Minehead, north Somerset. Corinna is a partner with Chris and with Jane Bassett in this venture, and owns a small but significant share of it. My greatest hero, the late Gerald Durrell (1925 – 1995) used to quip that *his* second wife Lee, "only married him for his zoo". There have been a lot of parallels between Durrell and myself. Indeed, just before Christmas last year, BBC4 broadcast a documentary about him. I was sitting in bed, unwell, watching it with Mark North sitting at the foot of my bed eating cake. As more and more details emerged about Durrell's views on life, and relationships with other people, Mark began to laugh so much that he nearly choked on cake crumbs. As I wrote in my autobiography, however, in every way Durrell dwarfs me. He was a giant of a man, both in his achievements and his failings. Whereas his achievements are so vast that they will forever leave mine looking insignificant in comparison, the same can be said about other aspects of his life. However, it is mildly amusing that where he claimed that his second wife married him for his zoo, I have married *my* second wife who is part-owner of one. There is a lot of work to be done with *Tropiquaria*, but their 2007 breeding successes with Jamaican boas, pancake tortoises, and northern helmeted curassows are something which can only be admired, and they are projects with which the CFZ is very proud to be involved.

Our publication schedule has continued apace this year. We are particularly proud of four books:

- *Monster: The A-Z Zooform Phenomena* by Neil Arnold
- *Big Cats: Loose in Britain* by Marcus Matthews
- *Man Monkey* by Nick Redfern
- *Extraordinary Animals Revisited* by Dr. Karl P.N. Shuker

The first two books have achieved a legendary status amongst the crypto-investigative community. Neil's has been in the offing for two or three years, and is the largest and most comprehensive run-down on zooform phenomena around the world that has ever been attempted. Marcus' book has an even longer history. It was written in the mid-1980s, and is the most comprehensive study of British big cats from pre-history to about 1990 ever to have been published. Researchers have known about this book for many years, and we are very proud to have at last been able to publish it.

Nick Redfern's book is the first volume ever to have been written on BHM phenomena in Britain, and Karl Shuker's book is the long-awaited and very welcome update to one of his best-loved and most obscure works. We are very proud to announce that we shall be publishing a whole string of books by both authors, which will include both updated and revised editions of their classic – and often long out of print – books, and entirely new volumes. Expect the next books by both of them in the New Year.

We are also very proud to be able to announce that the *CFZ Yearbook* series has been re-started. After a hiatus of three years, the *2007 Yearbook* was published this spring, and the *2008 Yearbook* – which will include this annual report – will be published within the next few weeks.

2007 also saw us publishing the second *Big Cat Yearbook* published by the Big Cats in Britain [BCIB] research group. This invaluable series includes in-depth reports on every known big cat sighting in the UK over a twelve-month period, together with essays from many of the luminaries in the field. It was particularly pleasing that the 2007 volume included a piece by Di Francis – one of the seminal researchers into big cat phenomena in the UK, and a researcher who has – sadly – fallen beneath the radar for the last few years.

For 2008 we are very pleased to announce yet another series. This will be of approximately 40 volumes and will be a county-by-county guide to the mystery animals, zooform phenomena, and animal folklore of the United Kingdom and the Republic of Ireland. The first volume which covers Tyneside and Northumberland has been written by Mike Hallowell and will be published early in the New Year. Other volumes have been commissioned from Karl Shuker, Nick Redfern, Richard Freeman, Jon

McGowan, Neil Arnold, Garry Cunningham, Oll Lewis and myself, and several of these will be published in 2008.

This summer we also launched a magazine called *Exotic Pets*. This does exactly what it says on the tin, and is intended to foster responsible animal keeping. We believe that under the current political climate, it may not be long before all exotic pet keeping has been hounded out of existence by the Animal Rights lobby and this increasingly disagreeable government. We believe that this will be a very bad thing, because the work of the amateur naturalist is irreplaceably valuable. Remember, that Darwin, Mendel, Gosse, Wallace, and, of course, Gerald Durrell, were all, basically, amateur naturalists. In the broadest sense of the word, so are the CFZ. Amateur naturalists are those people who do what they do un-constrained by the forces of finance or peer pressure.

Successive governments culminating in the shower who are presently strutting along the corridors of power have inflicted immeasurable damage upon the British way of life. They have turned our television companies from being the best in the world; national treasures who produce the world's greatest documentaries, dramas, and current affairs programmes, into a pathetic, shambling affair whose only remit seems to be to provide 'bread and circuses' to an increasingly moronic viewing public. They have turned our education system into a fatuous joke, and they have instituted such a draconian 'nanny state' that most of the things that once made Britain great are now nought but a vaguely embarrassing memory. We feel that one of the major roles of the CFZ in the 21st Century is to enthuse successive generations into rediscovering the joy of discovery of the natural world which my generation had as children. Natural history was a national obsession for about a century from the 1850s onwards, and had it not been for this national obsession, much of what we know about the natural world would still be a mystery. We believe that with publications such as *Exotic Pets* and *Animals & Men* that we are doing our own little bit to redress this balance, and we hope that you will agree with us that what we do is indeed a valuable thing.

This year's Weird Weekend, which took place on the third weekend in August, was a great success. It raised over £1,500 for CFZ funds, and attracted nearly 200 people from across the world. It was very encouraging to see how many of these were children who are becoming enthused by what we do. This year's speakers included Grigory Panchenko – the world's greatest expert on the almasty of the Caucasus – whose appearance at the Weird Weekend was his first speaking engagement in the West. Other speakers were Larry Warren, Peter Robbins, Richard Freeman, Oll Lewis, Nick Redfern, Jonathan McGowan, Matthew Williams, Mike Hallowell, Dr Charles Paxton, Adam Davies, Dr Darren Naish, Chris Moiser, Paul Vella, Ronan Coghlan and myself.

Next year's event will once again take place on the third weekend of August and confirmed speakers so far are: Adrian Shine, Mike Hallowell, Nick Redfern, Lee Walker, Matt Salusbury, Geoff Ward, Richard Freeman, Ronan Coghlan, Paul Vella and myself. Tickets will be going on sale in early January, and we strongly urge you to book your places early and also to book your accommodation in good time. Last year every B&B and campsite for miles was chocker.

- CFZ YEARBOOK 2008 -

In October this year we launched yet another new proejct. *'On the Track'* is amonthly web TV show hosted on our CFZtv website which provides an overview of cryptozoological news in general and CFZ news in particular. The third episode was posted in early December, and initial viewing figures of about 1,000 a month are very encouraging.

In November, for the fifth year in a row, the CFZ undertook a major foreign expedition. This year's excursion was to Guyana and was sponsored by Capcom. The five-person team consisted of:

- Richard Freeman (team leader)
- Dr. Chris Clark
- Jon Hare
- Lisa Dowley
- Paul Rose

Paul Rose is perhaps better known to his legions of fans as 'Mr. Biffo', the one-time head honcho of the Channel 4 teletext video game magazine *Digitiser*. He is better known these days as a humorous journalist, and TV script writer. Richard and I have been massive fans of his for years, and it was a great joy to us both, when, in May, he became a member of the CFZ. During the sojourn in the little explored grassland savannah of southern Guyana, he quickly became an integral part of the team, and whereas he intitially only went along because he had been commissioned to write a major book about the CFZ, he told me on the telephone the other night that he is now firmly 'hooked', and will be joining in CFZ activities for many a year to come.

The expedition went in search of information about three unknown animals:

- The didi (pronounced dai dai) – a bigfoot type creature.
- The giant anaconda
- The water tiger (a little known aquatic cryptid)

We found out a lot of information about all three species, but also came back with a wealth of data on two other cryptids as well. These are – as far as we are aware – completely unknown outside Guyana:

- A race of tiny red-faced pygmys
- A very small species of cayman

The team also secured the first ever video footage of a very recently discovered species of bright green scorpion, and were the first Europeans to visit some remote mountain caves which contained ancient burials.

Early in the New Year we will be publishing the expedition report written by all of the members of the team, and containing hundreds of unpublished photographs. We will also be releasing a full length documentary on CFZtv as part of our ongoing commitment to making the results of our investigations freely available to anyone who is interested.

Our other expedition this year was on a much smaller scale. However, in June, we returned to the Lake District, and in conjunction with our good friend Kevin Boyd, we spent four days in what turned out to be an ultimately fruitless hunt for eels. Back in 2002 we launched 'The Big Fish Project' which was initially a survey of extremely large fish from around the world, and the folklore surrounding it. However, in the last few years, this has morphed into what is perhaps the most important research project that we have ever done. With every new piece of data we become more convinced that the vast majority of so-called lake monsters from across the Northern Hemisphere are in fact enormous eels. Although the European eel (*Anguilla anguilla*) has been known for millenia, and has been a species of important economic concern for an equal time, surprisingly little is known about its biology. It was only well within the past 100 years that the details of its life-cycle were discovered, and various other important aspects of its biology are still being revealed. It was only in the past few years, for example, that it was discovered that a surprisingly last proportion of the European eel population stays offshore, and never actually enters freshwater, and it was only just over a year ago that it was discovered that the closely related species *Anguilla japonica* spawns and dies in the Marianas Trench near the Phillippines.

In our own little way, we have made a couple of potentially momentous discoveries within the field of eel biology. We have photographed eels far bigger than they are meant to grow, hidden away in the rather unpleasant intersteces of a run-down northern English holiday resort, and we have discovered historical evidence which suggests that the two currently known morphological varieties of the European eel are relatively recent, and appear only to have evolved in the past couple of hundred years.

Our investigations into the population dynamics of the European eels in lakes where monstrous specimens seem to have been reported will continue, and we hope that we shall eventually be successful in securing specimens, which we can take back and measure their growth rates under laboratory conditions.

Financially this has not been a good time for the CFZ. Every project that we have listed in this annual report has come in significantly over budget. Whereas we are very pleased with having secured the sponsorship deals described above, and the two smaller sponsorship deals from the *Farmer's Arms* in Woolsery, and the Cairngorm Brewery Company, for the Weird Weekend, we are disappointed that our other requests for sponsorship have been fruitless. Running the CFZ is an increasingly expensive business, and whilst none of the directors are paid anything for their time and effort, we have increasingly been forced to draw upon our own private resources in order to keep the organisation afloat. I have romped through the money I inherited from my late father, and there is very little left.

-CFZ YEARBOOK 2008-

Historically, we have funded the CFZ by doing contract design work, website building, and freelance journalism for a number of clients. During my father's final illness two years ago, we were forced to cut back on outside work, and by this summer we only had one regular client left. We severed ties with this client after an unfortunate and highly embarrassing incident which took place over the Weird Weekend. None of it was our fault, and everybody who knows the details has been appalled and disgusted by our ex-client's behaviour. However, this means that the only income we have comes from occasional articles for *Fortean Times* or *Beyond*, and sales of our own publications. The CFZ is in the worst financial position in which it has found itself since 2001, and with our new premises in the North Devon countryside, our overheads are higher than ever. We would be very grateful for any fund-raising ideas, potential sponsorship concepts, or any other ways that we can – as the late Sir John Verney once said – *"keep both ends of the wolf from meeting at the door"*.

However, all in all it has been a fantastic year. We shall draw a discreet curtain over the unfortunate events alluded to above which left us £500 a month worse off, and with our late client still owing us £2,500. We shall also draw a discreet curtain over the disgusting behaviour of NatWest Bank in closing our bank accounts for no valid reason, and we will do our best to forget the car crash which nearly killed Corinna and myself in September. But all in all, from where I am sitting, the future looks pretty rosy.

Thank you for your support over the last twelve months. Together we will continue to prove the truth of Bernard Heuvelmans' famous axiom that *"the great days of zoology are not done"*.

Here's to the future.

Jon Downes,
Director, CFZ
Woolsery, North Devon

3rd December 2007

THE CENTRE FOR FORTEAN ZOOLOGY

So, what is the Centre for Fortean Zoology?

We are a non profit-making organisation founded in 1992 with the aim of being a clearing house for information, and coordinating research into mystery animals around the world. We also study out of place animals, rare and aberrant animal behaviour, and Zooform Phenomena; – little-understood "things" that appear to be animals, but which are in fact nothing of the sort, and not even alive (at least in the way we understand the term).

Why should I join the Centre for Fortean Zoology?

Not only are we the biggest organisation of our type in the world but - or so we like to think - we are the best. We are certainly the only truly global Cryptozoological research organisation, and we carry out our investigations using a strictly scientific set of guidelines. We are expanding all the time and looking to recruit new members to help us in our research into mysterious animals and strange creatures across the globe. Why should you join us? Because, if you are genuinely interested in trying to solve the last great mysteries of Mother Nature, there is nobody better than us with whom to do it.

What do I get if I join the Centre for Fortean Zoology?

For £12 a year, you get a four-issue subscription to our journal *Animals & Men*. Each issue contains 60 pages packed with news, articles, letters, research papers, field reports, and even a gossip column! The magazine is A5 in format with a full colour cover. You also have access to one of the world's largest collections of resource material dealing with cryptozoology and allied disciplines, and people from the CFZ membership regularly take part in fieldwork and expeditions around the world.

How is the Centre for Fortean Zoology organized?

The CFZ is managed by a three-man board of trustees, with a non-profit making trust registered with HM Government Stamp Office. The board of trustees is supported by a Permanent Directorate of full and part-time staff, and advised by a Consultancy Board of specialists - many of whom who are world-renowned experts in their particular field. We have regional representatives across the UK, the USA, and many other parts of the world, and are affiliated with other organisations whose aims and protocols mirror our own.

I am new to the subject, and although I am interested I have little practical knowledge. I don't want to feel out of my depth. What should I do?

Don't worry. We were *all* beginners once. You'll find that the people at the CFZ are friendly and approachable. We have a thriving forum on the website which is the hub of an ever-growing electronic community. You will soon find your feet. Many members of the CFZ Permanent Directorate started off as ordinary members, and now work full time chasing monsters around the world.

I have an idea for a project which isn't on your website. What do I do?

Write to us, e-mail us, or telephone us. The list of future projects on the website is not exhaustive. If you have a good idea for an investigation, please tell us. We may well be able to help.

How do I go on an expedition?

We are always looking for volunteers to join us. If you see a project that interests you, do not hesitate to get in touch with us. Under certain circumstances we can help provide funding for your trip. If you look on the future projects section of the website, you can see some of the projects that we have pencilled in for the next few years.

In 2003 and 2004 we sent three-man expeditions to Sumatra looking for Orang-Pendek - a semi-legendary bipedal ape. The same three went to Mongolia in 2005. All three members started off merely subscribers to the CFZ magazine.

Next time it could be you!

Project Kerinci, Sumatra - 2003
In search of the bipedal ape Orang Pendek

How is the Centre for Fortean Zoology funded?

We have no magic sources of income. All our funds come from donations, membership fees, works that we do for TV, radio or magazines, and sales of our publications and merchandise. We are always looking for corporate sponsorship, and other sources of revenue. If you have any ideas for fund-raising please let us know. However, unlike other cryptozoological organisations in the past, we do not live in an intellectual ivory tower. We are not afraid to get our hands dirty, and furthermore we are not one of those organisations where the membership have to raise money so that a privileged few can go on expensive foreign trips. Our research teams both in the UK and abroad, consist of a mixture of experienced and inexperienced personnel. We are truly a community, and work on the premise that the benefits of CFZ membership are open to all.

What do you do with the data you gather from your investigations and expeditions?

Reports of our investigations are published on our website as soon as they are available. Preliminary reports are posted within days of the project finishing.

Each year we publish a 200 page yearbook containing research papers and expedition reports too long to be printed in the journal. We freely circulate our information to anybody who asks for it.

Is the CFZ community purely an electronic one?

No. Each year since 2000 we have held our annual convention - the *Weird Weekend* - in Exeter. It is three days of lectures, workshops, and excursions. But most importantly it is a chance for members of the CFZ to meet each other, and to talk with the members of the permanent directorate in a relaxed and informal setting and preferably with a pint of beer in one hand. Starting this year-18-20 August 2006 - the *Weird Weekend* will be bigger and better and held in the idyllic rural location of Woolsery in North Devon.

We are hoping to start up some regional groups in both the UK and the US which will have regular meetings, work together on research projects, and maybe have a mini convention of their own.

Since relocating to North Devon in 2005 we have become ever more closely involved with other community organisations, and we hope that this trend will continue. We also work closely with Police Forces across the UK as consultants for animal mutilation cases, and during 2006 we intend to forge closer links with the coastguard and other community services. We want to work closely with those who regularly travel into the Bristol Channel, so that if the recent trend of exotic animal visitors to our coastal waters continues, we can be out there as soon as possible.

We are building a Visitor's Centre in rural North Devon. This will not be open to the general public, but will provide a museum, a library and an educational resource for our members (currently over 400) across the globe. We are also planning a youth organisation which will involve children and young people in our activities.

Apart from having been the only Fortean Zoological organisation in the world to have consistently published material on all aspects of the subject for over a decade, we have achieved the following concrete results:

- Disproved the myth relating to the headless so-called sea-serpent carcass of Durgan beach in Cornwall 1975
- Disproved the story of the 1988 puma skull of Lustleigh Cleave
- Carried out the only in-depth research ever into mythos of the Cornish Owlma
- Made the first records of a tropical species of lamprey
- Made the first records of a luminous cave gnat larva in Thailand.
- Discovered a possible new species of British mammal - The Beech Marten.
- In 1994-6 carried out the first archival fortean zoological survey of Hong Kong.
- In the year 2000, CFZ theories where confirmed when an entirely new species of lizard was found resident in Britain.
- Identified the monster of Martin Mere in Lancashire as a giant wels catfish
- Expanded the known range of Armitage's skink in the Gambia by 80%
- Obtained photographic evidence of the remains of Europe's largest known pike
- Carried out the first ever in-depth study of the *ninki-nanka*
- Carried out the first attempt to breed Puerto Rican cave snails in captivity
- Were the first European explorers to visit the `lost valley` in Sumatra

EXPEDITIONS & INVESTIGATIOINS TO DATE INCLUDE

- 1998 Puerto Rico, Florida, Mexico *(Chupacabras)*
- 1999 Nevada *(Bigfoot)*
- 2000 Thailand *(Giant snakes called nagas)*
- 2002 Martin Mere *(Giant catfish)*
- 2002 Cleveland *(Wallaby mutilation)*
- 2003 Bolam Lake *(BHM Reports)*
- 2003 Sumatra *(Orang Pendek)*
- 2003 Texas *(Bigfoot; giant snapping turtles)*
- 2004 Sumatra *(Orang Pendek; cigau, a sabre-toothed cat)*
- 2004 Illinois *(Black panthers; cicada swarm)*
- 2004 Texas *(Mystery blue dog)*
- 2004 Puerto Rico *(Chupacabras; carnivorous cave snails)*
- 2005 Belize *(Affiliate expedition for hairy dwarfs)*
- 2005 Mongolia *(Allghoi Khorkhoi aka Mongolian death worm)*
- 2006 Gambia *(Gambo - Gambian sea monster , Ninki Nanka and Armitage s skink*
- 2006 Llangorse Lake *(Giant pike, giant eels)*
- 2006 Windermere *(Giant eels)*
- 2007 Coniston Water *(Giant eels)*
- 2007 Guyana *(Giant anaconda, didi, water tiger)*

To apply for a <u>FREE</u> information pack about the organisation and details of how to join, plus information on current and future projects, expeditions and events.

Send a stamped and addressed envelope to:

**THE CENTRE FOR FORTEAN ZOOLOGY
MYRTLE COTTAGE, WOOLSERY,
BIDEFORD, NORTH DEVON
EX39 5QR.**

or alternatively visit our website at:
www.cfz.org.uk

Other books available from
CFZ PRESS

THE OWLMAN AND OTHERS - 30th Anniversary Edition
Jonathan Downes - ISBN 978-1-905723-02-7

£14.99

EASTER 1976 - Two young girls playing in the churchyard of Mawnan Old Church in southern Cornwall were frightened by what they described as a "nasty bird-man". A series of sightings that has continued to the present day. These grotesque and frightening episodes have fascinated researchers for three decades now, and one man has spent years collecting all the available evidence into a book. To mark the 30th anniversary of these sightings, Jonathan Downes has published a special edition of his book.

DRAGONS - More than a myth?
Richard Freeman - ISBN 0-9512872-9-X

£14.99

First scientific look at dragons since 1884. It looks at dragon legends worldwide, and examines modern sightings of dragon-like creatures, as well as some of the more esoteric theories surrounding dragonkind.

Dragons are discussed from a folkloric, historical and cryptozoological perspective, and Richard Freeman concludes that: "When your parents told you that dragons don't exist - they lied!"

MONSTER HUNTER
Jonathan Downes - ISBN 0-9512872-7-3

£14.99

Jonathan Downes' long-awaited autobiography, *Monster Hunter*...

Written with refreshing candour, it is the extraordinary story of an extraordinary life, in which the author crosses paths with wizards, rock stars, terrorists, and a bewildering array of mythical and not so mythical monsters, and still just about manages to emerge with his sanity intact.......

MONSTER OF THE MERE
Jonathan Downes - ISBN 0-9512872-2-2

£12.50

It all starts on Valentine's Day 2002 when a Lancashire newspaper announces that "Something" has been attacking swans at a nature reserve in Lancashire. Eyewitnesses have reported that a giant unknown creature has been dragging fully grown swans beneath the water at Martin Mere. An intrepid team from the Exeter based Centre for Fortean Zoology, led by the author, make two trips – each of a week – to the lake and its surrounding marshlands. During their investigations they uncover a thrilling and complex web of historical fact and fancy, quasi Fortean occurrences, strange animals and even human sacrifice.

CFZ PRESS, MYRTLE COTTAGE, WOOLFARDISWORTHY BIDEFORD, NORTH DEVON, EX39 5QR
w w w . c f z . o r g . u k

Other books available from
CFZ PRESS

ONLY FOOLS AND GOATSUCKERS
Jonathan Downes - ISBN 0-9512872-3-0

£12.50

In January and February 1998 Jonathan Downes and Graham Inglis of the Centre for Fortean Zoology spent three and a half weeks in Puerto Rico, Mexico and Florida, accompanied by a film crew from UK Channel 4 TV. Their aim was to make a documentary about the terrifying chupacabra - a vampiric creature that exists somewhere in the grey area between folklore and reality. This remarkable book tells the gripping, sometimes scary, and often hilariously funny story of how the boys from the CFZ did their best to subvert the medium of contemporary TV documentary making and actually do their job.

WHILE THE CAT'S AWAY
Chris Moiser - ISBN: 0-9512872-1-4

£7.99

Over the past thirty years or so there have been numerous sightings of large exotic cats, including black leopards, pumas and lynx, in the South West of England. Former Rhodesian soldier Sam McCall moved to North Devon and became a farmer and pub owner when Rhodesia became Zimbabwe in 1980. Over the years despite many of his pub regulars having seen the "Beast of Exmoor" Sam wasn't at all sure that it existed. Then a series of happenings made him change his mind. Chris Moiser—a zoologist—is well known for his research into the mystery cats of the westcountry. This is his first novel.

CFZ EXPEDITION REPORT 2006 - GAMBIA
ISBN 1905723032

£12.50

In July 2006, The J.T.Downes memorial Gambia Expedition - a six-person team - Chris Moiser, Richard Freeman, Chris Clarke, Oll Lewis, Lisa Dowley and Suzi Marsh went to the Gambia, West Africa. They went in search of a dragon-like creature, known to the natives as `Ninki Nanka`, which has terrorized the tiny African state for generations, and has reportedly killed people as recently as the 1990s. They also went to dig up part of a beach where an amateur naturalist claims to have buried the carcass of a mysterious fifteen foot sea monster named 'Gambo', and they sought to find the Armitage's Skink (*Chalcides armitagei*) - a tiny lizard first described in 1922 and only rediscovered in 1989. Here, for the first time, is their story.... With an forward by Dr. Karl Shuker and introduction by Jonathan Downes.

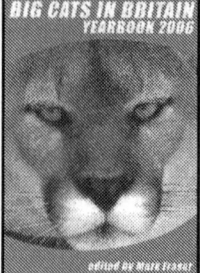

BIG CATS IN BRITAIN YEARBOOK 2006
Edited by Mark Fraser - ISBN 978-1905723-01-0

£10.00

Big cats are said to roam the British Isles and Ireland even now as you are sitting and reading this. People from all walks of life encounter these mysterious felines on a daily basis in every nook and cranny of these two countries. Most are jet-black, some are white, some are brown, in fact big cats of every description and colour are seen by some unsuspecting person while on his or her daily business. 'Big Cats in Britain' are the largest and most active group in the British Isles and Ireland This is their first book. It contains a run-down of every known big cat sighting in the UK during 2005, together with essays by various luminaries of the British big cat research community which place the phenomenon into scientific, cultural, and historical perspective.

CFZ PRESS, MYRTLE COTTAGE, WOOLSERY, BIDEFORD, NORTH DEVON, EX39 5QR
w w w . c f z . o r g . u k

Other books available from
CFZ PRESS

THE SMALLER MYSTERY CARNIVORES OF THE WESTCOUNTRY
Jonathan Downes - ISBN 978-1-905723-05-8

£7.99

Although much has been written in recent years about the mystery big cats which have been reported stalking Westcountry moorlands, little has been written on the subject of the smaller British mystery carnivores. This unique book redresses the balance and examines the current status in the Westcountry of three species thought to be extinct: the Wildcat, the Pine Marten and the Polecat, finding that the truth is far more exciting than the currently held scientific dogma. This book also uncovers evidence suggesting that even more exotic species of small mammal may lurk hitherto unsuspected in the countryside of Devon, Cornwall, Somerset and Dorset.

THE BLACKDOWN MYSTERY
Jonathan Downes - ISBN 978-1-905723-00-3

£7.99

Intrepid members of the CFZ are up to the challenge, and manage to entangle themselves thoroughly in the bizarre trappings of this case. This is the soft underbelly of ufology, rife with unsavoury characters, plenty of drugs and booze." That sums it up quite well, we think. A new edition of the classic 1999 book by legendary fortean author Jonathan Downes. In this remarkable book, Jon weaves a complex tale of conspiracy, anti-conspiracy, quasi-conspiracy and downright lies surrounding an air-crash and alleged UFO incident in Somerset during 1996. However the story is much stranger than that. This excellent and amusing book lifts the lid off much of contemporary forteana and explains far more than it initially promises.

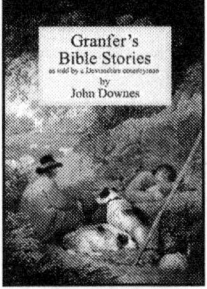

GRANFER'S BIBLE STORIES
John Downes - ISBN 0-9512872-8-1

£7.99

Bible stories in the Devonshire vernacular, each story being told by an old Devon Grandfather - 'Granfer'. These stories are now collected together in a remarkable book presenting selected parts of the Bible as one more-or-less continuous tale in short 'bite sized' stories intended for dipping into or even for bed-time reading. `Granfer` treats the biblical characters as if they were simple country folk living in the next village. Many of the stories are treated with a degree of bucolic humour and kindly irreverence, which not only gives the reader an opportunity to re-evaluate familiar tales in a new light, but do so in both an entertaining and a spiritually uplifting manner.

FRAGRANT HARBOURS DISTANT RIVERS
John Downes - ISBN 0-9512872-5-7

£12.50

Many excellent books have been written about Africa during the second half of the 19th Century, but this one is unique in that it presents the stories of a dozen different people, whose interlinked lives and achievements have as many nuances as any contemporary soap opera. It explains how the events in China and Hong Kong which surrounded the Opium Wars, intimately effected the events in Africa which take up the majority of this book. The author served in the Colonial Service in Nigeria and Hong Kong, during which he found himself following in the footsteps of one of the main characters in this book; Frederick Lugard – the architect of modern Nigeria.

**CFZ PRESS, MYRTLE COTTAGE,
WOOLFARDISWORTHY BIDEFORD,
NORTH DEVON, EX39 5QR
www.cfz.org.uk**

Other books available from
CFZ PRESS

ANIMALS & MEN - Issues 1 - 5 - In the Beginning
Edited by Jonathan Downes - ISBN 0-9512872-6-5

£12.50

At the beginning of the 21st Century monsters still roam the remote, and sometimes not so remote, corners of our planet. It is our job to search for them. The Centre for Fortean Zoology [CFZ] is the only professional, scientific and full-time organisation in the world dedicated to cryptozoology - the study of unknown animals. Since 1992 the CFZ has carried out an unparalleled programme of research and investigation all over the world. We have carried out expeditions to Sumatra (2003 and 2004), Mongolia (2005), Puerto Rico (1998 and 2004), Mexico (1998), Thailand (2000), Florida (1998), Nevada (1999 and 2003), Texas (2003 and 2004), and Illinois (2004). An introductory essay by Jonathan Downes, notes putting each issue into a historical perspective, and a history of the CFZ.

ANIMALS & MEN - Issues 6 - 10 - The Number of the Beast
Edited by Jonathan Downes - ISBN 978-1-905723-06-5

£12.50

At the beginning of the 21st Century monsters still roam the remote, and sometimes not so remote, corners of our planet. It is our job to search for them. The Centre for Fortean Zoology [CFZ] is the only professional, scientific and full-time organisation in the world dedicated to cryptozoology - the study of unknown animals. Since 1992 the CFZ has carried out an unparalleled programme of research and investigation all over the world. We have carried out expeditions to Sumatra (2003 and 2004), Mongolia (2005), Puerto Rico (1998 and 2004), Mexico (1998), Thailand (2000), Florida (1998), Nevada (1999 and 2003), Texas (2003 and 2004), and Illinois (2004). Preface by Mark North and an introductory essay by Jonathan Downes, notes putting each issue into a historical perspective, and a history of the CFZ.

BIG BIRD! Modern Sightings of Flying Monsters

£7.99

Ken Gerhard - ISBN 978-1-905723-08-9

From all over the dusty U.S./Mexican border come hair-raising stories of modern day encounters with winged monsters of immense size and terrifying appearance. Further field sightings of similar creatures are recorded from all around the globe. What lies behind these weird tales? Ken Gerhard is a native Texan, he lives in the homeland of the monster some call 'Big Bird'. Ken's scholarly work is the first of its kind. On the track of the monster, Ken uncovers cases of animal mutilations, attacks on humans and mounting evidence of a stunning zoological discovery ignored by mainstream science. Keep watching the skies!

STRENGTH THROUGH KOI
They saved Hitler's Koi and other stories

£7.99

Jonathan Downes - ISBN 978-1-905723-04-1

Strength through Koi is a book of short stories - some of them true, some of them less so - by noted cryptozoologist and raconteur Jonathan Downes. The stories are all about koi carp, and their interaction with bigfoot, UFOs, and Nazis. Even the late George Harrison makes an appearance. Very funny in parts, this book is highly recommended for anyone with even a passing interest in aquaculture, but should be taken definitely *cum grano salis*.

**CFZ PRESS, MYRTLE COTTAGE,
WOOLSERY, BIDEFORD,
NORTH DEVON, EX39 5QR**

Other books available from
CFZ PRESS

CFZ PRESS

BIG CATS IN BRITAIN YEARBOOK 2007
Edited by Mark Fraser - ISBN 978-1-905723-09-6

£12.50

People from all walks of life encounter mysterious felids on a daily basis, in every nook and cranny of the UK. Most are jet-black, some are white, some are brown; big cats of every description and colour are seen by some unsuspecting person while on his or her daily business. 'Big Cats in Britain' are the largest and most active research group in the British Isles and Ireland. This book contains a run-down of every known big cat sighting in the UK during 2006, together with essays by various luminaries of the British big cat research community.

CAT FLAPS! Northern Mystery Cats
Andy Roberts - ISBN 978-1-905723-11-9

£6.99

Of all Britain's mystery beasts, the alien big cats are the most renowned. In recent years the notoriety of these uncatchable, out-of-place predators have eclipsed even the Loch Ness Monster. They slink from the shadows to terrorise a community, and then, as often as not, vanish like ghosts. But now film, photographs, livestock kills, and paw prints show that we can no longer deny the existence of these once-legendary beasts. Here then is a case-study, a true lost classic of Fortean research by one of the country's most respected researchers.

CENTRE FOR FORTEAN ZOOLOGY 2007 YEARBOOK
Edited by Jonathan Downes and Richard Freeman
ISBN 978-1-905723-14-0

£12.50

The Centre For Fortean Zoology Yearbook is a collection of papers and essays too long and detailed for publication in the CFZ Journal *Animals & Men*. With contributions from both well-known researchers, and relative newcomers to the field, the Yearbook provides a forum where new theories can be expounded, and work on little-known cryptids discussed.

MONSTER! THE A-Z OF ZOOFORM PHENOMENA
Neil Arnold - ISBN 978-1-905723-10-2

£14.99

Zooform Phenomena are the most elusive, and least understood, mystery `animals`. Indeed, they are not animals at all, and are not even animate in the accepted terms of the word. Author and researcher Neil Arnold is to be commended for a groundbreaking piece of work, and has provided the world's first alphabetical listing of zooforms from around the world.

**CFZ PRESS, MYRTLE COTTAGE,
WOOLFARDISWORTHY BIDEFORD,
NORTH DEVON, EX39 5QR
w w w . c f z . o r g . u k**

Other books available from
CFZ PRESS

BIG CATS LOOSE IN BRITAIN
Marcus Matthews - ISBN 978-1-905723-12-6

£14.99

Big Cats: Loose in Britain, looks at the body of anecdotal evidence for such creatures: sightings, livestock kills, paw-prints and photographs, and seeks to determine underlying commonalities and threads of evidence. These two strands are repeatedly woven together into a highly readable, yet scientifically compelling, overview of the big cat phenomenon in Britain.

DARK DORSET
TALES OF MYSTERY, WONDER AND TERROR
Robert. J. Newland and Mark. J. North
ISBN 978-1-905723-15-6

£12.50

This extensively illustrated compendium has over 400 tales and references, making this book by far one of the best in its field. Dark Dorset has been thoroughly researched, and includes many new entries and up to date information never before published. The title of the book speaks for itself, and is indeed not for the faint hearted or those easily shocked.

MAN-MONKEY - IN SEARCH OF THE BRITISH BIGFOOT
Nick Redfern - ISBN 978-1-905723-16-4

£9.99

In her 1883 book, *Shropshire Folklore*, Charlotte S. Burne wrote: *'Just before he reached the canal bridge, a strange black creature with great white eyes sprang out of the plantation by the roadside and alighted on his horse's back'*. The creature duly became known as the `Man-Monkey`.

Between 1986 and early 2001, Nick Redfern delved deeply into the mystery of the strange creature of that dark stretch of canal. Now, published for the very first time, are Nick's original interview notes, his files and discoveries; as well as his theories pertaining to what lies at the heart of this diabolical legend.

EXTRAORDINARY ANIMALS REVISITED
Dr Karl Shuker - ISBN 978-1905723171

£14.99

This delightful book is the long-awaited, greatly-expanded new edition of one of Dr Karl Shuker's much-loved early volumes, *Extraordinary Animals Worldwide*. It is a fascinating celebration of what used to be called romantic natural history, examining a dazzling diversity of animal anomalies, creatures of cryptozoology, and all manner of other thought-provoking zoological revelations and continuing controversies down through the ages of wildlife discovery.

**CFZ PRESS, MYRTLE COTTAGE,
WOOLFARDISWORTHY BIDEFORD,
NORTH DEVON, EX39 5QR
www.cfz.org.uk**

Other books available from
CFZ PRESS

DARK DORSET CALENDAR CUSTOMS
Robert J Newland - ISBN 978-1-905723-18-8

£12.50

Much of the intrinsic charm of Dorset folklore is owed to the importance of folk customs. Today only a small amount of these curious and occasionally eccentric customs have survived, while those that still continue have, for many of us, lost their original significance. Why do we eat pancakes on Shrove Tuesday? Why do children dance around the maypole on May Day? Why do we carve pumpkin lanterns at Hallowe'en? All the answers are here! Robert has made an in-depth study of the Dorset country calendar identifying the major feast-days, holidays and celebrations when traditionally such folk customs are practiced.

CENTRE FOR FORTEAN ZOOLOGY 2004 YEARBOOK
Edited by Jonathan Downes and Richard Freeman
ISBN 978-1905723140

£12.50

The Centre For Fortean Zoology Yearbook is a collection of papers and essays too long and detailed for publication in the CFZ Journal *Animals & Men*. With contributions from both well-known researchers, and relative newcomers to the field, the Yearbook provides a forum where new theories can be expounded, and work on little-known cryptids discussed.

**CFZ PRESS, MYRTLE COTTAGE,
WOOLFARDISWORTHY BIDEFORD,
NORTH DEVON, EX39 5QR
w w w . c f z . o r g . u k**

www.ingramcontent.com/pod-product-compliance
Lightning Source LLC
Chambersburg PA
CBHW060655100426
42734CB00047B/1886